SpringerBriefs in Applied Sciences and Technology

PoliMI SpringerBriefs

More information about this subseries at https://link.springer.com/bookseries/11159
http://www.polimi.it

Manuela Grecchi

Building Renovation

How to Retrofit and Reuse Existing Buildings to Save Energy and Respond to New Needs

POLITECNICO
MILANO 1863

Springer

Manuela Grecchi
Dipartimento ABC
Politecnico di Milano
Milan, Italy

ISSN 2191-530X ISSN 2191-5318 (electronic)
SpringerBriefs in Applied Sciences and Technology
ISSN 2282-2577 ISSN 2282-2585 (electronic)
PoliMI SpringerBriefs
ISBN 978-3-030-89835-9 ISBN 978-3-030-89836-6 (eBook)
https://doi.org/10.1007/978-3-030-89836-6

This Springer imprint is published by the registered company Springer Nature Switzerland AG
The registered company address is: Gewerbestrasse 11, 6330 Cham, Switzerland

Preface

The chapters in this book deal with building renovation and the resulting urban renewal.

The urgency of climate change and all the problems associated with the excessive use of energy are forcing a reorganisation of the process interested in reusing the existing building stock with a more sustainable approach.

The adaptive conversion of old buildings has become a dominant theme in many urban renewal projects. It must necessarily include strategies for energy efficiency and reduction of pollutant emissions, improvement of environmental performance, economic sustainability and cultural identity.

The topics covered in the chapters are the result of various research projects on building renovation and adaptive reuse, as well as the refurbishment course that the author has been teaching for more than 10 years.

The book aims to review the process of building renovation by systematising the phases of analysis and prior knowledge with a project that not only pays attention to energy saving, but also thoroughly examines complex issues, such as defining correct new functions and answers to new needs.

The examples selected in this book are intended to provide evidence of good practice in the review and transformation of old buildings, with two main focuses that represent critical situations in many cities: the conversion of former industrial areas to initiate urban renewals, and the adaptation to new needs and high construction performance of existing residential districts with critical social problems.

Milan, Italy

Manuela Grecchi

Contents

Chapter 1
Building Renovation Opportunity

Abstract The urgency of climate change and all the problems related to the excessive use of energy force a reorganisation of the process which is interested in the reuse of the existing building stock. Building refurbishment and adaptive reuse are often confusing terms, but in reality they point to approaches that focus on different objectives. The need to systematise decisions and evaluate options at different levels of intervention highlights the importance of codifying processes and adopting effective decision models. Different environmental, economic and social sustainability challenges need to be taken into account along with how building refurbishment could provide a solution.

Keywords Building renovation · Adaptive reuse · Building obsolescence

1.1 Introduction

Buildings are designed for a certain lifetime, at the end of which a decision has to be made about their future. Expectations for the lifetime of a building are typically 50–60 years (Chapman et al., 2002). After decades of use, a building may no longer meet the needs of its occupants.

The built environment is often at the centre of discussions on combating climate change. Construction activities are responsible for more than 40% of total energy consumption and an important part of greenhouse gas and carbon dioxide emissions. Considering that the European building stock is renovated at a rate of only 1–1.5% per year, there is considerable scope for reducing energy consumption by working on existing buildings.

To reduce overall energy consumption, renovation is often preferable to demolition and new construction. Studies have shown that the real life of buildings exceeds their estimated useful life and, in terms of the impact on resources and the environment, renovation is preferable to demolition. The built environment is an important part of the European cultural heritage and a source of socio-economic development (Thuvander et al., 2012).

The renovation process is generally more complex than for a new building in terms of decision-making, planning and execution. The characteristics, quality and

M. Grecchi, *Building Renovation*, PoliMI SpringerBriefs,
https://doi.org/10.1007/978-3-030-89836-6_1

weaknesses of the existing building must be known. There is a risk that architectural, cultural and social values may be overlooked in favour of energy and economic performance parameters (Thuvander et al., 2012).

In order to make the right choices, supportive tools are needed to ensure an appropriate balance between wishes, needs and values and a number of important aspects such as energy, environment, technical and economic performance, and social, cultural and architectural aspects.

1.1.1 Terminology

There is not a single definition to describe building changes, but several terms are used, often overlapping. Common terms include adaptation, renovation, refurbishment, reconstruction, retrofitting, restoration, reconstruction and transformation. The use of different terminology is due to the typological and dimensional diversity of buildings, the wide range of measures and the variety of motivations determining an intervention, such as conservation, technical, functional or social obsolescence (Thuvander et al., 2012).

While some terms are used with the same meaning, such as conservation, preservation or restoration, other terms have a wider scope of interpretation. In the scientific literature, some authors use the term renovation to denote a minimum number of interventions, while others find that the term refers to a more substantial rehabilitation. The term retrofitting is often used to emphasise the act of raising the standard of a building, for example in terms of energy efficiency (Thuvander et al., 2012).

Changes to buildings can vary from minor repairs or adaptations with limited intervention to major renovations with substantial transformations. The range of works extends from protection or maintenance work limited to stopping the deterioration, usually to preserve the original building, to renovation or replacement by demolition and reconstruction.

Modifications include both new additions and those which significantly alter part or all of the existing building. The change is not defined by the extent of the intervention, but by its consequences. In terms of regulations, the modified building is subject to the same technical and energy efficiency requirements as a new building.

Adaptation is an expression that has been widely interpreted and defined by many researchers. The term is derived from the Latin 'ad' (to) 'aptare' (adapt). Typically, the definitions refer to "change of use", highest "preservation" of the original structure and layout of a building and the extension of its "useful life" (Wilkinson et al., 2014).

In this book, "renovation" stands for any intervention to adapt and improve a building to new conditions or requirements. Adaptive reuse is assumed to focus on a renovation that involves a change of use.

The aim is to achieve a sustainable renovation that takes into account efforts to meet the requirements of environmental, social and economic sustainability in the conversion of buildings.

The term "sustainable renovation" is used to emphasise a holistic approach that balances environmental, social and economic aspects.

There are different definitions depending on the context in which the term sustainability is used. It is broader than "meeting the needs of the present without compromising the ability of future generations to meet their own needs".[1]

1.2 The Need for Building Renovation

It is estimated that people worldwide spend a total of about 70% of their time in buildings, which explains why healthy and efficient constructions should be a priority.

The need to renovate buildings can also be related to indoor environmental quality issues. It has long been generally accepted that a poor indoor climate can lead to health problems. The combination of several disciplines adds complexity to building renovation, so interdisciplinary expertise is required.

Today the focus is mainly on energy efficiency and deterioration, but there are also non-energy benefits of building refurbishment, such as improved environmental and lighting performance, better floor plans, safety, accessibility, etc. (Jensen & Maslesa, 2015).

The renovation of existing buildings can contribute to a more sustainable built environment. Various environmental, economic and social sustainability challenges need to be evaluated along with how building renovation could provide a solution.

Environmental problems—Buildings cause energy waste, greenhouse gas emissions, water and resource consumption. As only 1–2% of new buildings are added each year, much of the built environment that will exist in 2050 is already built. Therefore, existing buildings need to be adapted quickly to comply with international climate agreements.

However, there are some positive signals. For example, in many developed countries more is being spent on adapting buildings than on constructing new buildings. In general, the renovation of buildings is more environmentally sustainable than demolition and reconstruction because it involves:

- less production of demolition waste,
- less consumption of resources,
- less grey energy during construction,
- less energy consumption,
- less pollution.

Economic problems—There are strong economic arguments in favour of building renovation, even if it is not economically advantageous but is balanced by social and environmental benefits. Convincing economic arguments can be the lower costs of a renovation intervention compared to the demolition and reconstruction of a building, the shorter construction time and the reduction of the demolition time.

[1] World Commission on Environment and Development, Brundtland Report 1987.

Not all projects are economically viable, especially in the case of limited or poorly constructed buildings. Nevertheless, costs are still a major adjustment factor.

Social problems—Finally, building refurbishment can contribute to social sustainability. A key argument is that the reuse of buildings enables society to preserve the social and cultural capital that buildings represent. More generally, renovation can be part of urban renewal processes. At the building level, renovation can increase user satisfaction. As users' needs are constantly changing, it is still necessary to take measures to extend the functional life of existing buildings. In other words, the renovation and reuse of buildings is inherently sustainable from a social perspective, as it contributes to the maintenance of:

- a feeling of place and belonging,
- interaction and user satisfaction.

1.2.1 Building Obsolescence

The built environment is constantly subject to changes and cycles of renewal. Some old buildings are demolished and replaced by new buildings, while others are preserved and adapted to new uses. What deserves to be renovated or restored? Is there a way to assess the limits of a transformation?

Obsolescence is perceived as a problem of economic and social degradation. Obsolescence can affect any building at any point in its life and can provide an opportunity for reuse and renovation. This depends first and foremost on the life span of the building.

Decisions on whether to upgrade an existing building are usually taken when its performance is no longer at the required level. Different types of obsolescence can affect buildings, and not only the physical condition must be considered. They are:

- ***Physical obsolescence***—This typically occurs when maintenance and repairs are insufficient. It can be measured in terms of the maintenance schedule and performance.
- ***Functional obsolescence***—The propensity for adaptability can be measured by determining the degree of flexibility integrated into the building configuration. The life span is effectively shortened if the layout of the buildings does not respond flexibly to changes.
- ***Technical obsolescence***—Performance, comfort and energy consumption are parameters that are evaluated to measure technological obsolescence. The lifetime is shortened considerably when occupant comfort depends on high energy consumption.
- ***Economic obsolescence***—Location, the relationship between the function of the building and the market, the changing value of the land due to urban development, commercial traffic and the market are all factors that can lead to economic obsolescence in an urban context. Economic obsolescence can be seen in the location of the building relative to current property values.

The reuse of an existing construction can involve considerable work. This can mean alterations, extensions, renovations or any other kind of intervention, and adaptive reuse often leads to the regeneration of an entire area. Many factors are relevant when making decisions about existing buildings. These factors are not mutually exclusive.

Economic advantages—There are a number of advantages that can encourage reuse:

- *Construction costs.* This is often the case when the cost of converting a building is less than the cost of demolishing and rebuilding it.
- *Shorter development time.* The total period for a project involving work on an existing building may be shorter than the period required for a new building.
- *Use of existing infrastructure.* When existing buildings are reused, the infrastructure may already exist, although an upgrade may be necessary.
- *Plot*[2] *ratios.* Many local regulations specify land parcel ratios for new developments.

Cultural heritage—Historical buildings are an important testimony to the culture of a country. There is a moral obligation to preserve them for future generations. Many historic buildings are representative of local materials and traditions.

History—Old buildings bear witness to the past in terms of many different aspects, such as culture, technologies, materials and typologies. Historical buildings provide a link between the past and the present. Knowledge of the past helps to understand the present and plan for the future.

Architecture—Some buildings are the products of outstanding architects. Others are links in the chain of architectural development, or they are milestones in the local context.

Quality—One of the main reasons for reusing old buildings is simply their quality. It can be related to architectural style, materials, typology and spatial organisation or performance levels. Some buildings can be outstanding architectural masterpieces.

Sustainability—Buildings are responsible for more than 40% of total energy consumption. Half of it is consumed during construction (grey energy) and half during the life of the building. If existing buildings are reused, significant energy savings can be achieved (e.g., less grey energy and less waste from demolition).

Disadvantages of reusing old buildings—Some of the main disadvantages are:

- *Functional.* An adapted building simply may not be able to meet the requirements in terms of performance, functional layout, floor to ceiling height and efficient use of the land.
- *Technical.* Existing structures may contain some latent defects that are too difficult or too expensive to solve.
- *Economic.* Old buildings are often more expensive to maintain than new ones.

---

[2] The term 'plot ratio' refers to the ratio between the total floor area of a building and the area of the site (or plot) on which the building stands.

Many other changes can occur over the course of a building's lifetime and affect a building. For example:

- changes in legal requirements (e.g., in relation to health and safety, accessibility for disabled people, etc.),
- new residents with different expectations,
- new needs of the existing residents,
- new technologies that affect the way buildings are used,
- social and economic changes can make the current use of a building obsolete.

When a building reaches a point below the minimum acceptable performance standard, an important decision must be made. There are essentially three possible decisions:

- sell the building and its area and identify an alternative building,
- demolish the building and construct a new building in its place,
- perform work to bring the building to a completely new level of performance.

A building has several different types of "life"; it can be considered as a cyclical process. During its lifetime, its use alternates with renovations. At a certain point in time, the building will reach a state where it is necessary to evaluate its performance, usability, and future value and calculate the obsolescence rate. This may be due to the technical or functional characteristics of the building (technical or functional obsolescence) or to the fact that the costs of use outweigh the benefits of employment (economic obsolescence). At this point, the building may undergo major adaptations or its useful life may come to an end. (Wilkinson et al., 2014).

Technical life span—The technical life span is the period of time during which a building fulfils the necessary technical and physical requirements. The technical performance of the building may be reduced due to the following factors:

- external or climatic conditions (sun, wind, rain),
- material ageing (chemical processes),
- use (loss of quality),
- regulations (restrictions or updates),
- wishes of owners and users (less acceptance and satisfaction).

The decay can be stopped or slowed down by proper maintenance and repairs (Fig. 1.1). Maintenance refers to repairs that ensure or restore the original functionality of the building, but not to the improvement of the original technical quality. If the technical quality falls below an acceptable level, stronger measures than routine maintenance are required. However, the technical life of a building ends when the owner is no longer willing to carry out the necessary repairs. This moment is often related to functional or economic circumstances.

Functional life span—Functional life is the period of time during which a building meets the needs of the occupants for all activities performed. The functional performance may decrease due to the following factors:

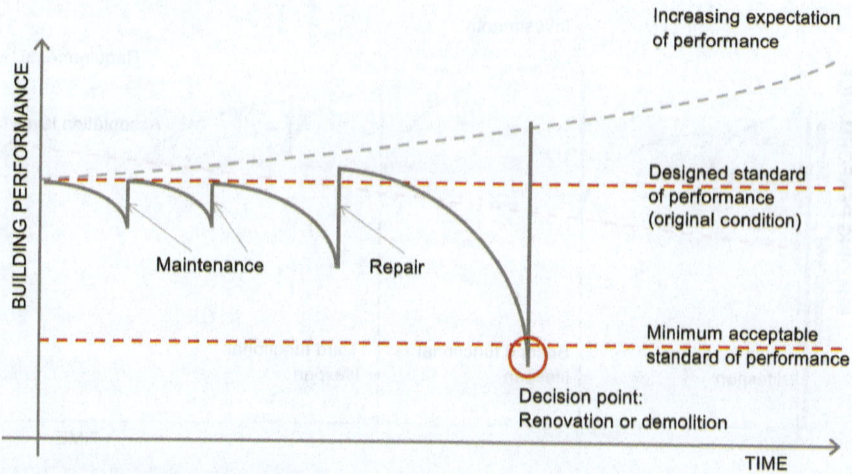

Fig. 1.1 Building technical life span

- regulations (stricter requirements for spatial dimensions),
- wishes of owners and users (appearance, size, level of services, accessibility, parking).

The functional life span ends when the building restricts the user in the exercise of his activities, which means that the functional quality of buildings is no longer accepted. In this case, various options for increasing the functional quality can be considered, for example, by changing the building (floor plan, façades, volume) or by changing its purpose. In the latter case, the demolition of the building should be mentioned.

Economic commitment is an important differentiating factor in the final decision (Fig. 1.2).

Economic life span—The economic life is the period during which the building generates more income than costs. The income that a building can generate is based on price, quality and market demand. Costs relate to maintenance and building management. The combined revenues and costs produce a cash flow for a given period. The economic performance of a building may decrease due to the following factors:

- general economic situation (a general economic recession leads to lower demand for space/buildings and lower rents),
- legislation (e.g., energy performance requirements),
- expectations of investors and users (higher returns for investors; lower costs for users).

The economic life of a building ends when the owner no longer expects to generate more revenue than costs, that is, if the cost–benefit ratio is (permanently) negative or if the building has a negative value (Fig. 1.3).

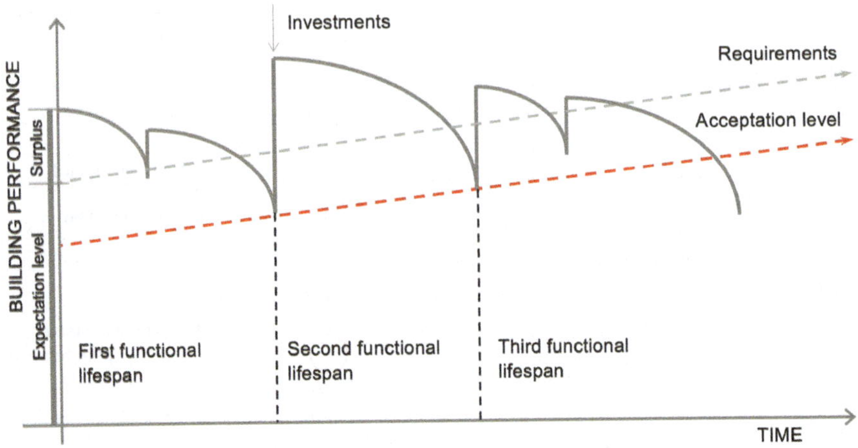

Fig. 1.2 Building functional life span

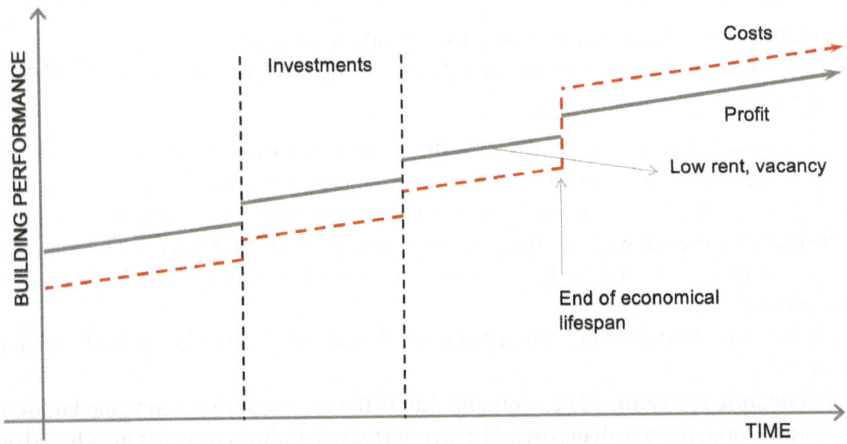

Fig. 1.3 Building economic life span

1.3 Conclusion

Building renovation and adaptive reuse are topics that require important evaluation before a project can begin. As said before, the process can be complex and a multi-disciplinary approach and structured support for decision-making are essential to combine all possible implications and consequences at different levels.

Energy is certainly a priority, but other aspects relating to user needs, local, environmental, social and economic opportunities must also be considered and arranged in the best possible balance to maximise the outcome of the building refurbishment in accordance with the context.

The next chapters will focus on specific issues, but it is crucial to keep an eye on the process in its complexity. Each renovation project has its own specificity and complexity, which often leads to solutions that cannot be replicated. For this reason, a clear understanding of the process and its necessary decision-making steps helps to keep the progress of the project under control.

References

Chapman, T. J. P., Chow, F. C., & Skinner, H. (2002). Building on old foundations—Sustainable construction for urban regeneration. In *CE-world conference, ASCE online conference*. ASCE, 2007.

Jensen, P. A., & Maslesa, E. (2015). Value based building renovation—A tool for decision-making and evaluation. *Building and Environment, 92*. (Elsevier).

Thuvander, L., Femenías, P., Mjörnell, K, & Meiling, P. (2012). Unveiling the process of sustainable renovation. *Sustainability, 4*(6).

Wilkinson, S., Remøy, H., & Langston, C., (2014). *Sustainable building adaptation. Innovation in decision-making*. Wiley Blackwell Editor.

Chapter 2
Building Renovation Process

Abstract To properly plan for building reuse, it is necessary to identify a process capable of driving all subsequent actions to achieve the best possible solution. The first fundamental step is a thorough preliminary study of the building and its context to gather various types of information. The more comprehensive this phase is, the more correct the subsequent decision phases will be. Multicriteria techniques are a suitable tool to select one or more reuse alternatives from a range of options that can be expected to provide reliable solutions.

Keywords Decision-making process · Preliminary investigation · Multi-criteria techniques

2.1 Introduction

The reuse and renovation of existing buildings means, above all, defining strategies for modernising buildings in line with new needs, performance requirements and user expectations in terms of well-being and control of comfort management systems.

The heterogeneity of buildings produced with significantly different technologies, often linked to local traditions and materials, and whose performance does not conform to current standards, underlines the urgent need not to impose standardised solutions but to define a process linking all the typical phases of reuse interventions.

A structural redesign makes sense if the chosen retrofit is in line with technical possibilities, economic demand, environmental conditions and social interests. It is permissible and justified if (Antucheviciene & Zavadskas, 2008):

- the building is structurally sound, reusable and suitable for the proposed use,
- the shape, size and general design of the building and the proposed transformation are in harmony with the environment, architectural styles and local materials,
- the proposed use does not harm the local environment by any form of pollution of water, soil or air.

Multi-criteria techniques are an appropriate means of selecting one or more alternatives from a set of available options, based on a range of often conflicting uses, in terms of several properly weighted criteria (Fig. 2.1).

M. Grecchi, *Building Renovation*, PoliMI SpringerBriefs,
https://doi.org/10.1007/978-3-030-89836-6_2

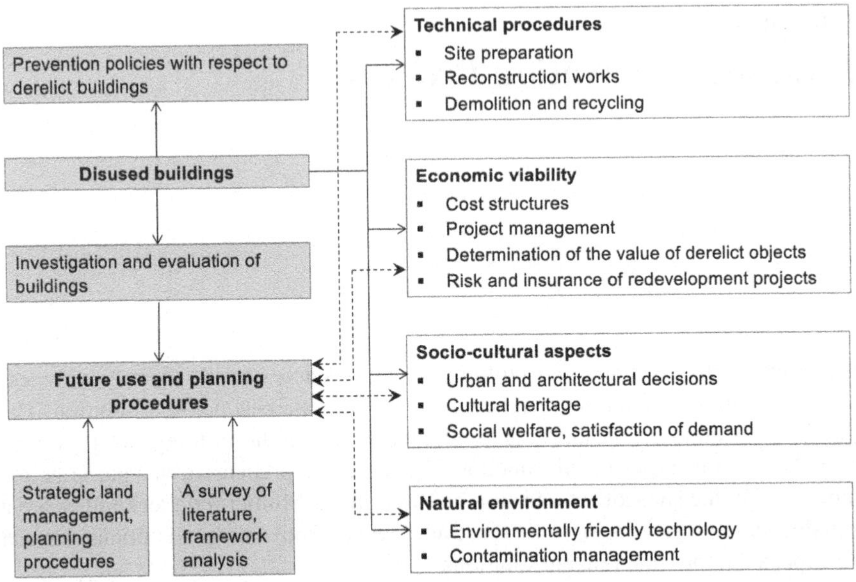

Fig. 2.1 Scheme of different stages of the process. *Source* Antucheviciene and Zavadskas (2008), revised by the author

The morphology and characteristics of a building, such as size and layout, number of floors, internal partitions and openings, structural condition, etc., have a major impact on the success of any reuse project. However, the assessment should also take into account external factors such as location, landscape quality, accessibility, available services and historical value.

In general, a renovation process has more or less the same phases as a new construction project (surveying, planning, design, construction, operation, use). However, in order to achieve good results, more attention must be paid to the preliminary study phase. In this context, it should be added that a renovation process often starts with poor documentation of the building.

In order to properly define the hypotheses for reuse, it is necessary to link and take into account all the data collected, to check their impact on the context and on what the solution will bring in terms of redesign.

2.2 Preliminary Investigation

An in-depth study is a broad term that involves the collection of several sets of information and an inventory of the characteristics and state of conservation of the building before the decision to renovate is taken.

The preliminary investigation must include surveys of the building and its surroundings (local context, technical condition and maintenance), documentation of user needs, property data, history (year of construction, previous renovations/alterations, users), drawings and images (construction drawings, historical drawings, sketches, photographs). Furthermore, a preliminary survey should be objective and not contain subjective assessments (Thuvander et al., 2012).

2.2.1 Data Collection

All surveys should be planned in relation to the project objectives. The information collected should not be an end in itself but, on the contrary, should be part of a linear process. It is important how the collected data is returned: all material must be systematically well organised to support the next steps.

A gradual transition from the context and general characteristics of the building to smaller architectural details is the right way to avoid mistakes and omissions.

The following images show some possible organisations of the collected data on a building scale. They were created by students in several years of the course *Refurbishment and Energy Retrofit*, of which the author is professor.

Documentation—Documentation is the basis of every study and analysis. Historical documents, old pictures, local books, maps and urban development can be helpful in various types of analysis. Bibliographic research is also one of the first measures to be considered.

Photographic overview—Photographs are the first step; they will be the basis for further research. A photogrammetric survey or rectified photography in combination with computer drawings can be a reliable method of capturing facades and interior details. Photographs can be used to determine a certain degree of decay of the building.

Context—Systematic and critical analyses at territorial level are necessary. Topography, landscape features, type of settlement, distance from the city centre, roads, connections and traffic, accessibility for pedestrians and vehicles, available services, existing functions and programs at urban scale, green and public spaces, historical buildings and monuments, social and local traditions, development of the territory and its main interesting transformations are the typical aspects that need to be known before a hypothesis of a possible reuse of the building can be made.

The analysis should also be extended to climate data and demographic trends that are relevant in terms of the number, composition and age of the population, occupations or traditions of the region.

All these analyses will show the "strengths" and "weaknesses" of the area and suggest possible new uses that can become important inputs for the project.

History—Knowledge of how the building has changed over the course of its life is fundamental. The investigation of significant changes, such as additions or demolitions, changes in use, and the dating of the various parts of the building, could

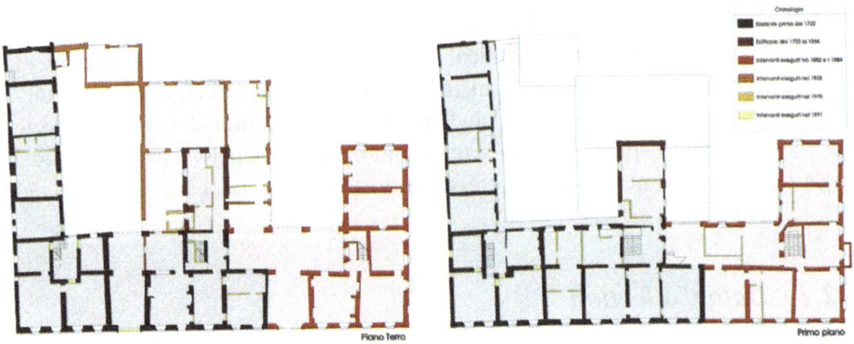

Fig. 2.2 Historical evolution, example of data representation

allow assumptions to be made at a later date about building technologies, causes of structural failure, or typical deterioration (Fig. 2.2).

General building characteristics—The shape, floor plan configuration, number of storeys, building system, interior partitions, roof and façade elements, and openings are general features that will form the background of the future project. Drawings of plans, sections and elevations form the basis for all further steps. Surveys must be accurate, all details are recorded, and comments on materials, finishes or other features are recorded on special cards linked to the drawings by codes.

Original uses—Knowledge of the original use and its changes over time, in relation to buildings or individual rooms, can be useful for further evaluation. The changed use may have led to significant changes in layout, addition of systems or the overloading of structures (Fig. 2.3).

Architecture—The analysis of recurrent "typologies" relating to plans and technical elements, architectural details and cultural heritage could form the basis for future decisions on what to transmit as a memory or to redesign during the project of adaptive reuse. Again, the notes and associated drawings may highlight information that can guide future design decisions.

Construction technologies—They represent one of the most important and delicate surveys necessary to understand problems and plan decisions.

Without knowledge of the original construction techniques, it is not possible to eliminate deterioration, repair structures, improve performance or add new layers. In order to obtain reliable information about the construction method, it may be necessary to conduct instrumental investigations or partial demolitions. However, correct assumptions can also be made by considering the date of construction and by comparing some details with typical solutions available in local manuals. In the past, each region has produced specific solutions using local materials; and a careful bibliographic search could help to compare traditional techniques with the details of the building. (See Chap. 5 for some insights and examples).

Deterioration—Every building gets old, but when ageing occurs too quickly or external phenomena cause deterioration, appropriate interventions are required, which are certainly different from normal maintenance. All investigations related to

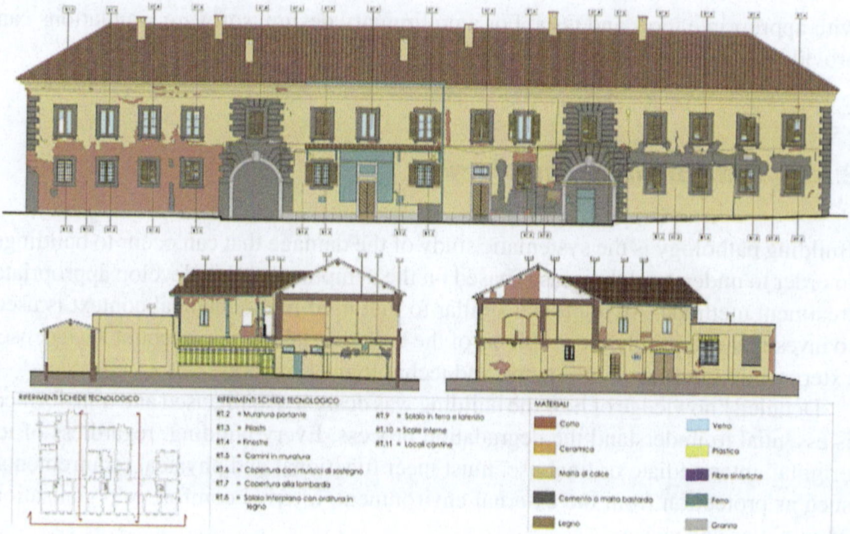

Fig. 2.3 Building technologies. The codes refer to specific cards

deterioration must be well organised before defining causes and remedies. Typically, "anomaly" cards (each card reports only one anomaly with specific information) and mapping boards (necessary to record the locations, extent, and overlap of deterioration) are organised with graphic codes. Only by crossing the two different scales of information, supported by the literature on the mechanisms of degradation, is it possible to make a definite diagnosis (Fig. 2.4).

Performance Analysis—Any building envelope, even if damaged, still provides performance even if it is insufficient to meet the required standards. Starting from a detailed technological section, it is possible to analyse the thermal behaviour, waterproofing, acoustics and load-bearing capacity. These findings are obtained

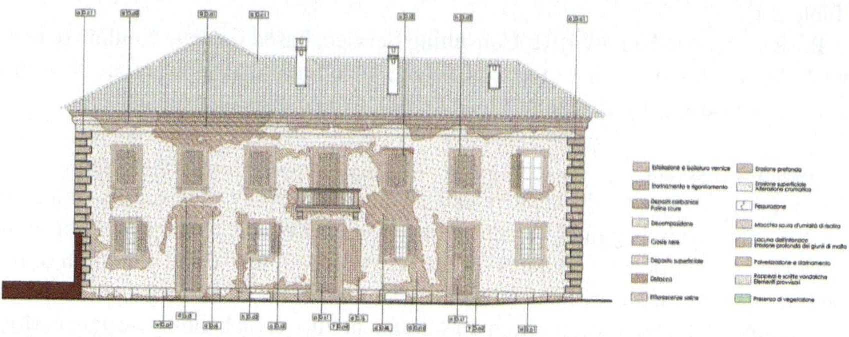

Fig. 2.4 Deterioration map. The codes refer to specific cards

with appropriate tools and tests. For a preliminary design, software simulations can provide reliable assumptions.

2.2.2 Box: Building Pathology

Building pathology is the systematic study of the damage that can occur to buildings in order to understand the causes based on the symptoms and to develop appropriate treatment methods. An approach similar to that used in the medical context is used to investigate and study the problems of the building, taking into account its age, use, external and internal environment, and technological features.

Detailed knowledge of how the building was designed, built, used and transformed is essential to understand the degradation process. Every building, regardless of its original, intermediate or final use, must meet functional and physical requirements, such as protection from the external environment, internal comfort or organisation of activities and space.

The causes of deterioration and decay of building materials are influenced by external and internal environmental parameters such as water, humidity, temperature, UV rays, light and lack of ventilation. The analysis of building pathologies is a complex task; it requires an integrated multidisciplinary approach combining materials science, physics, chemistry, environmental monitoring and conservation sciences. Historical buildings are constructed according to traditional methods using various combinations of local materials.

Damage and Decay

The different elements of a building are subject to different forms of defects or failure. Many problems are related to the poor quality of design, construction, repair and maintenance.

The classification of impacts and deterioration mechanisms that affect buildings (ISO 6241/1984) provides useful information for identifying the causes of problems (Table 2.1).

Research carried out by BRE Consulting Service, based on their building defects database, identified ten main types of defect. Damp and excess water are the most common causes of deterioration.

Causes and Effects

It is important to properly identify and evaluate the decay mechanism acting on the building before beginning repair works. The effects and consequences of these mechanisms vary depending on the construction, location, use and condition of the building and the effects of local climatic and atmospheric characteristics.

Climate—Climate variations have a significant impact on building design, performance and durability of materials. The natural and built environment has an impact on local climatic conditions. Microclimates are influenced by factors such as topography, vegetation, buildings and human activities.

Table 2.1 Classification of agents working outside and inside buildings, ISO 6241/1984, Watt (2007)

Agents	Acting outside the building		Acting inside the building	
	Atmosphere	Soil	Occupancy	Design consequences
Mechanical agents				
Gravitation	Snow and rainwater loads	Soil and water pressure	Live loads	Dead loads
Forces and imposes or restrained deformations	Ice formation pressure, thermal and moisture expansion	Subsidence, slip	Handling forces, indentation	Shrinkage, creep, forces and imposed deformations
Kinetic energy	Wind, hail, external impacts, sandstorm	Earthquakes	Internal impacts, wear	Water hammer
Vibration and noises	Wind, thunder, airplanes, explosions, traffic, machinery noises	Traffic and machinery vibration	Noise and vibration from music, dancers, domestic appliances	Services noises and vibrations
Electromagnetic agents				
Radiation	Solar radiation, radioactive radiation	Radioactive radiation	Lamps, radioactive radiation	Radiating surfaces
Electricity	Lighting	Stray currents	–	Static electricity, electrical supply
Magnetism	–	–	Magnetic fields	Magnetic fields
Thermal agents				
Thermal agents	Solar radiation, radioactive radiation	Radioactive radiation	Lamps, radioactive radiation	Radiating surfaces
	Heat, frost, thermal shock	Soil heat, frost	User-emitted heat, cigarette	Heating, fire
Chemical agents				
Water and solvents	Air humidity, condensations, precipitations	Surface and soil water	Water spray, condensation, detergents, alcohol	Water supply, wastewater seepage
Oxidizing agents	Oxygen, ozone, nitrous, oxides	Positive electrochemical potentials	Disinfectant, bleach	Positive electromechanical potentials

(continued)

Table 2.1 (continued)

Agents	Acting outside the building		Acting inside the building	
	Atmosphere	Soil	Occupancy	Design consequences
Reducing agents	–	Sulphides	Agents of combustion, ammonia	Agents of combustion, negative electrochemical potentials
Acids	Carbonic acid, bird droppings, sulphuric acid	Carbonic acid, humic acids	Vinegar, citric acid, carbonic acid	Sulphuric acid, carbonic acid
Bases	–	Lime	Sodium, potassium, ammonium	Sodium hydroxide, cement
Salts	Salty fog	Nitrates, phosphates, chlorides, sulphate	Sodium chloride	Calcium chloride, sulphate, plaster
Chemically neutral	Neutral dust	Limestone, silica	Fat, oil, ink, neutral dust	Fat, oil, ink, neutral dust
Biological agents				
Vegetable and microbial	Bacteria, seeds	Bacteria, mold, fungi, roots	Bacteria, house plants	–
Animal	Insects, birds	Rodents, termites, worms	Domestic animals	–

Weathering—Sun, wind and rain cause changes in building materials. Mechanical and physical effects are caused by frost and extreme temperature fluctuations; water absorbed by porous materials creates pressure on the internal structure when it freezes (water expands when it freezes), causing cracks.

A chemical process can produce carbonation, oxidation and hydration. Chemical and mechanical processes can occur because of plants and animals.

Light and other electromagnetic radiation—These can cause photochemical reactions. Ultraviolet (UV) radiation can alter the electronic distribution in materials and cause colour changes, oxidation or embrittlement.

Atmospheric gases and pollutants—Water vapour, pollutants and particles can have a dangerous effect on materials and people. Pollutants are typically caused by industrial emissions, traffic and the burning of fossil fuels. Sea salt and other chlorides can damage porous materials by crystallising within the pore structure.

Chemical, Physical and Biological Effects

Chemical, physical, and biological processes stress materials in interaction with the environment and other components. Many of these processes are complex and require specialised knowledge to identify causes and find the best solutions.

Table 2.2 Natural and anthropogenic risk factors affecting buildings, Watt (2007)

Natural risk factors	Nature	Timing	Scale	Probability
Biological factors				
Animals, insects, fungi, plants	Biological	Long-term	Local/regional	Continuous
Meteorological factors				
Solar radiation, precipitation, frost	Physical	Long-term	Local	Continuous cycles
Air humidity and temperature	Physical	Long-term	Local	Continuous cycles
Lighting	Physical	Rapid onset	Local	Infrequent
Gale force winds, tornadoes, hurricanes	Physical	Rapid onset	Regional	Rare/infrequent
Geochemical hazards				
Groundwater erosion	Physiochemical	Long-term	Local	Continuous
Soluble salts	Physiochemical	Long-term	Local	Continuous
Natural hazards				
Soil movement	Physical	Long-term	Regional	Continuous
Sea surge	Physical	Rapid onset	Local	Frequent
River and sea floods	Physical	Rapid onset	Local/regional	Rare/infrequent
Earthquakes	Physical	Rapid onset	Local/regional	Rare/infrequent
Fires	Physical	Rapid onset	Local	Rare
Landslides, avalanches	Physical	Rapid onset	Local	Rare
Volcanic eruptions	Physical	Rapid onset	Local/regional	Rare
Anthropogenic risk factors	*Nature*	*Timing*	*Scale*	*Probability*
Pollution				
Air pollution	Chemical	Long-term	Local/regional/global	Continuous

(continued)

Table 2.2 (continued)

Natural risk factors	Nature	Timing	Scale	Probability
Acid rain	Chemical	Long-term	Regional/global	Continuous
Mass tourism				
Mass tourism	Physical/biological/chemical	Long-term	Local	Continuous
Management				
Heating, air conditioning, ventilation, lighting	Physical/chemical	Long-term	Local	Continuous
Inappropriate standards of comfort	Physical/chemical	Long-term	Regional/local	Continuous
Inappropriate use				
Occasional use with sudden warming	Physical	Long-term	Local	Continuous
Localized vibration	Physical	Long-term	Local	Continuous
Transformation of buildings	Architectural	Rapid onset	Local	Frequent
Bad care				
Improper cleaning	Physical/chemical	Long-term	Local	Continuous
Late intervention	Physical/biological/chemical	Medium-term	Local	Continuous
Careless handling, neglect	Physical/biological/chemical	Medium-term	Local	Infrequent
Inadequate remedial actions	Physical/biological/chemical	Medium-term	Local	Infrequent
Remedial works				
Bad choice of experts	Physical/biological/chemical	Rapid onset	Local	Infrequent
Treatments	Physical/chemical	Long-term	Local	Infrequent
Improper repair	Physical/chemical	Long-term	Local	Infrequent
Inappropriate restoration	Physical/chemical	Long-term	Local	Infrequent
Human hazards				

(continued)

Table 2.2 (continued)

Natural risk factors	Nature	Timing	Scale	Probability
Economic development	Environmental	Medium-term	Regional/global	Continuous
Governmental policy	No limits	Rapid onset/medium-term	Regional/global	Infrequent/continuous
Accidents, theft, vandalism	Physical/chemical	Rapid onset	Local	Infrequent
Arson, terrorism	Physical	Rapid onset	Local	Infrequent
War	Physical	Rapid onset	Regional/global	Infrequent

Chemical actions—Materials are constantly exposed to the effects of humans and the environment. Many of these interactions lead to chemical reactions that often cause irreversible changes. Typical examples are metal corrosion, sulphate attacks in cement-based mortars and carbonation of concrete.

Physical actions—Materials stressed by heat, moisture, crystallisation of soluble salts, light, sound, electricity, and magnetism may undergo a physical change that alters their performance.

Biological actions—Vegetation can damage buildings by direct contact with walls and roofs or underground by the action of roots on foundations and drains. Climbing plants can cause damage through aerial roots, suckers, tendrils and the secretion of acidic substances (Fig. 2.5).

Movements

Building components and materials are subject to movement due to temperature and moisture fluctuations, interactions between the building load and the substrate, or other external influences on the load-bearing capacity of the building. In order to design appropriate remedial measures, it is important to understand the origin of building movements.

Human Factors

Human activities have a significant impact on the performance of buildings. During their lifetime, materials, equipment and all components are exposed to aggressive conditions. Improper repairs and maintenance can lead to irreversible damage to the building.

Transformation and change of use can cause different overloads, vibrations, physical changes and different environmental conditions.

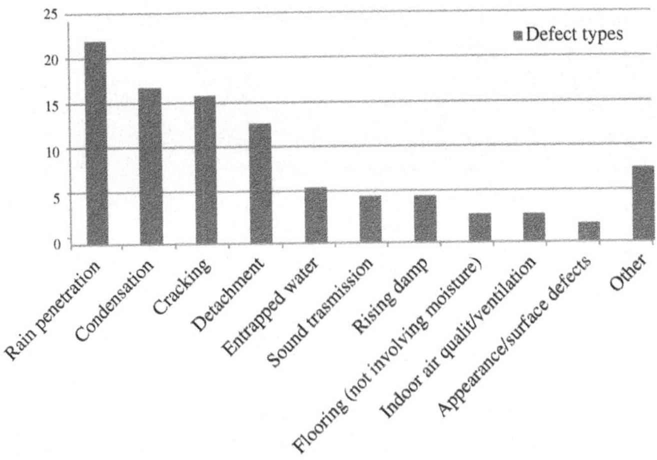

Fig. 2.5 Attribution of defect types (BRE Advisory Service)

Diagnostics

As Harris (2001) notes, deterioration mechanisms are nonlinear but multivariable functions. Accurate simulations of the phenomena are often difficult to compute, and when the building is viewed as a series of interconnected subsystems, the mechanisms of simultaneous deterioration are particularly complex to analyse.

Building Inspections and Surveys

To understand the causes of degradation, inspections and investigations are often necessary. Detailed knowledge of the building and its history (date of construction, technologies, alterations and repairs) is essential to plan inspections and organise the following steps.

- ***Pre-visit on-site*** to inform the inspector of the scope and nature of the inspections; to familiarise themselves with the floor plan, residents and neighbours; to plan special equipment and inspection services; to take photos and sketches for upcoming visits.
- ***Background research***, including construction date, legal protection, previous uses, site characteristics, repair and maintenance activities over the years.
- ***Inspections and investigations***, covering a range of different activities, such as visual observations and instrumental studies.
- ***Report***, all relevant information gathered in the previous phases and the results of the investigations. It is a guide for upcoming decisions and actions.

Diagnosis of Defects

A logical set of questions and answers provides a method for identifying the process of decay and the causes associated with the defect. A "decision tree" then makes it possible to link causes and effects step by step and to relate the specific problem to all available information.

According to Watt, if the symptoms are particularly complex, a more scientific approach may be required. This approach is based on several steps:

- observation to identify visible symptoms,
- theories to develop various hypotheses based on the observations,
- questions to choose between theories,
- tests to find answers to questions through specialised surveys,
- analysis to interpret test results,
- conclusions to compare the diagnosis with the symptoms and results,
- action to program appropriate work.

Non-destructive Examination Techniques

Non-destructive methods can be used to investigate defects, providing information based on direct inspection. The term "non-destructive" means that the element is not damaged. Each technique is capable of producing a specific set of data on material properties, hidden anomalies, or for performance evaluation.

The diagnosis of certain defects may require monitoring over time using simple or sophisticated methods. Specific diagnostic tools such as moisture meters, wall surface thermometers and salt detectors can be used to determine the moisture, dew point temperature, risk of surface condensation and the presence of hygroscopic salts. Simple on-site tests can measure the moisture content or alkalinity of concrete. Chemical and physical tests can be important to identify materials and determine the degree of decay.

Thermography—The measurement of waves radiating from the surface is often used to detect and quantify heat loss and temperature variations. It can also be used to detect hidden details such as the detachment of panelling, embedded columns, woodwork and moisture phenomena.

Infrared photography—Photography with infrared film and suitable filters can show slight variations in surface colours caused by an increase in humidity or deterioration of the substrate.

Radiography—X-rays are used to detect material discontinuities, cracks, structural parts, and hidden details. Dense materials such as masonry and concrete require a more powerful radiation source.

Acoustic test—Various techniques use sound waves to detect inconsistencies within materials. By using transducers on both sides of the structure, it is possible to measure differences in time and frequency response. Ultrasonic techniques use high frequency sound waves to locate hidden details, voids, defects, cracks and cavities and measure thickness and density.

Geophysical surveying—Non-destructive geophysical methods can be used to identify underground features. Resistivity techniques analyse the resistance of an electric current flowing through the ground and detect fluctuations in this resistance, which may indicate trenches, walls and various soil features.

Sclerometer test—This technique measures the hardness of materials such as plaster and concrete. The sclerometer measures the surface hardness of the material and its resistance to pressure.

Magnetometry—A technique used to locate magnetic elements embedded in non-magnetic materials and commonly used to identify reinforcement in concrete structures, metal fittings and fixings in masonry.

Fibre optic inspection—A fibre optic probe can provide useful information about the condition of components. Endoscopes, which are flexible tubes with optical fibres that can illuminate and view distant objects, are useful for inspecting and photographing hidden details.

Carbonation test—By spraying a phenolphthalein indicator solution that reacts in an alkaline environment, this method tests the carbonation process within a concrete component. A concrete that has not been carbonated is stained purple.

Petrographic analysis—Various techniques are used to identify rock types and their conservation. Thin section analysis shows the sequence of rock layers with a detailed description of thickness, regularity, composition and degradation.

Monitoring of Defects

Many defects are progressive and need to be monitored over a period of time. Monitoring usually includes absolute and relative measures.

Environmental monitoring—Knowledge of the external and internal environmental parameters makes it possible to link assumptions about the causes of a defect to its specific context. These parameters include temperature (environment and surface), humidity (absolute and relative), sunlight (lux and ultraviolet), wind (speed and direction), air movement and pollution. It is important to take data variations into account.

Structural movement monitoring—Techniques may vary depending on the specific problem identified. Analysis of cracks and planar deformations in masonry can be performed using crack width gauges or rulers, calibrated indicator lights, vernier gauges and fixed bolts, mechanical strain gauges or electronic sensors connected to data loggers.

The deformations of the components can be monitored with levelling devices, inclinometers or electronic level gauges; vibrations can be observed with laser interferometry or with sensors and data loggers; the deformation of vertical surfaces can be monitored with simple plummets, more precisely with an optical plummet. @ @ @

2.3 The Decision-Making Process for Renovation

As mentioned earlier, a renovation process is not far from the design of a new building. Figure 2.6 illustrates the general steps typically used for a reuse project. Note that the preliminary design and drafting phases are iterative and that "sub-iterations" occur throughout the process (Nielsen et al., 2016). There are six areas where methods of decision-making can contribute (Fig. 2.7).

Goal setting—The first fundamental step is the definition of goals and criteria, as all subsequent steps are adapted to these strategic and important aspects (Ferreira et al., 2013). The objectives can be based on some stakeholder evaluations, or the criteria can be defined from the beginning, leaving the weighting to the decision-maker.

Evaluation of the current state of the building—The evaluation after the definition of the project objectives will make the process more focused, but the diagnosis can be made before the objectives are set.

Design alternatives—Several options could be developed at this stage. Design alternatives are normally the responsibility of a design team and may therefore involve subjective judgements.

Performance evaluation of design alternatives—Simulation tools can be used, ranging from simple economic and energy calculations to complex and exhaustive simulations to assess the performance of the project.

Evaluation of project alternatives—In this phase, it is possible to evaluate quantitatively and qualitatively different renovation scenarios. If the goal is building certification, a sustainability assessment is carried out.

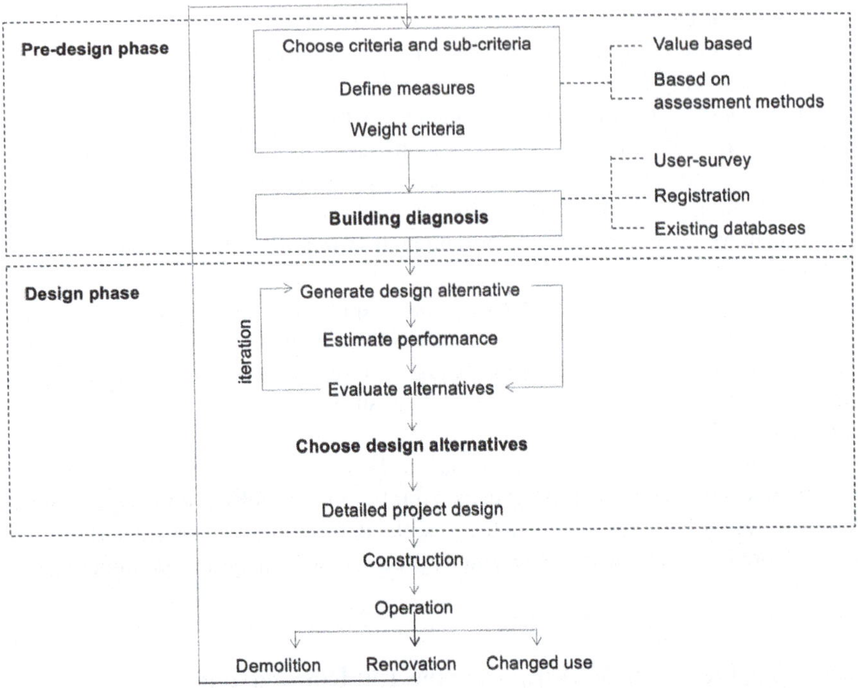

Fig. 2.6 Building renovation process. *Source* Nielsen et al. (2016), revised by the author

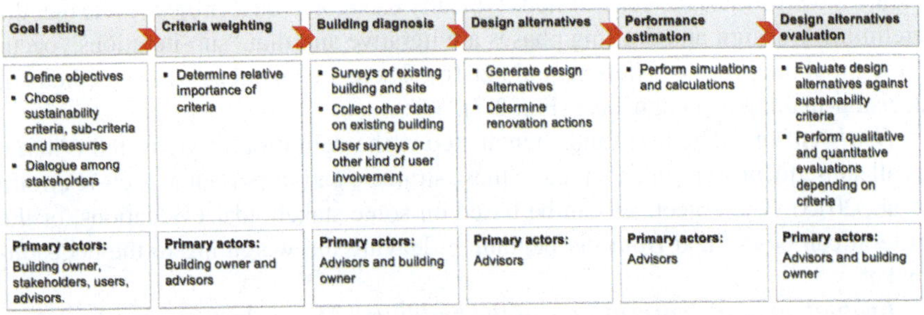

Fig. 2.7 Areas where decision making methods can assist renovation designs. *Source* Nielsen et al. (2016), revised by the author

The decision-making process for the renovation of buildings is complex. Many stakeholders are involved, each with a different perspective. Decision-makers are investors, developers, users and retailers. Another level of complexity is that these stakeholders make decisions at different stages of the process and that each of them has a different level of impact.

2.4 A Sustainable Approach

Green building and sustainable renovation are terms that are often used to briefly describe a design process that selects and links all decisions with the aim of saving energy and reducing air pollutants. The choice of materials, building technologies and what affects the well-being of the users will guide all phases.

"Green" and "sustainable" are terms that are often used synonymously to describe similar objectives, such as the use of natural light and ventilation, solar energy and non-toxic building materials. As already mentioned in the first chapter, sustainability has a broader meaning, as it considers not only buildings but also the environment and evaluates all aspects that arise from their use and the relationship between ecology, economy and social well-being.

The integrated design provides a process for achieving these goals. The specifics of a project may vary, but the means to achieve these goals must be consistent. Factors such as climate, the master plan, building shape, façades and thermophysical properties of materials must play a strategic role in project decisions. Integrated design is a circular process of collaboration where knowledge from different disciplines is brought together to find the best solution.

Energy—The plight of climate change is forcing solutions to reduce carbon emissions and policies to transform the built environment into low-carbon cities. Existing buildings are responsible for more than 40% of the global energy consumption, cause a third of greenhouse gas emissions and are concentrated in urban areas. By 2050, 70% of the world population will live in cities. Historic buildings and large suburban extensions built after World War II in response to immigration and social need, often with inferior materials and poor solutions, represent the majority of existing buildings where action needs to be taken to meet the required energy and emissions targets. According to Yung and Chan (2012), the renovation and reuse of buildings avoids the wasteful process of demolition and reconstruction. The environmental upgrading of existing buildings should focus on economic, environmental and social benefits.

Economical—Adapting a building can be cheaper than demolishing and rebuilding it. Before deciding what is the most cost-effective solution, a comparison between the two different approaches should be considered because, as Wilkinson (2012) noted, poor building quality could dramatically increase the cost of refurbishment, making the construction of a new building more feasible.

Environmentally friendly—Green retrofitting means less energy consumption. According to Ball (2002), environmental and social benefits can influence the decision to retrofit, even if the economic costs are high. The impact of land use on the natural environment is multifaceted, such as changes in habitats and ecosystems, which could lead to some problems in the balance between the built and natural environment.

Social and cultural issues—Greening historic buildings could be a challenge, considering that technical solutions must preserve architectural value while solving problems related to energy savings and carbon emissions. In addition, cultural heritage or monuments have a historical and social value for the local community.

Table 2.3 above briefly lists the criteria developed by Yung and Chan to introduce sustainable aspects into a reuse process.

Figure 2.8 summarises the key factors that influence the decision to renovate a building and implement an adaptive reuse process. The dashed line underlines the links between different factors.

2.4.1 Main Criteria for Environmental Design

Figure 2.9 shows the conceptual framework of a model developed by Chan and Lee (2009) for a Hong Kong renewal project. This model is a good example of how complex decisions on the adaptive reuse of large areas can be made, taking into account all aspects of sustainability.

The model defines a hierarchy consisting of three main levels:

1. Goal
2. Goal of sustainability
3. Design criteria.

The top level is the goal, the search for the most sustainable renewal solution. The second level is the sustainability goals, which take into account three different aspects:

1. Economic
2. Ecological
3. Social.

The third level consists of 17 design criteria extracted from 30 urban design project considerations prepared by Chan and Lee (2009). The six grey shaded criteria in the figure highlight the environmental aspects. Their relationship to environmental sustainability is briefly discussed below.

The links between criteria and environmental aspects are explained by Lee and Chan in this way:

- ***Access to work***—The location of the workplace has a major impact on environmental quality. If the workplace is outside the home, people tend to use their own car. If the workplace is easily accessible, people are more likely to reach it by public transport, bicycle or on foot.
- ***Community spirit***—People who have a strong sense of belonging to their community are more likely to develop a deep sense of responsibility towards their living environment. They are therefore more likely to protect urban ecology.
- ***Green design***—It is generally accepted that green design can optimise the use of the natural resources available in a community, improve the environmental quality of the city and minimise the waste of scarce resources. The use of natural resources can be more effective through good building design. The total amount of energy consumed and the rate of greenhouse gas emissions will decrease, making a significant contribution to curbing global warming.

Table 2.3 Significant sustainable adaptive reuse criteria identified in the literature and verified by project participants

Sustainable adaptive reuse criteria	
Criterion	Description
Economical	
Economic viability	Extent of financing itself through commercially viable activities
Job creation	Number of jobs created by the new use which contributes to the local economy
Tourism	Generate tourism industry and revenue in the area and the country
Cost efficiency	Renovation cost versus economic return generated by the new use
Compliance with statutory regulations	Extent of the adaptive reuse complies with regulations
Social and cultural	
Sense of place and identity	Feeling of attachment to the place and create a common cultural identity
Continuity of social life	Ensure the continuity of their life and the cultural significance of the place
Social cohesion and inclusiveness	Facilitating a fairer distribution of resources
Environmental and physical	
Environmental performance	Physical building performance, e.g. energy efficiency, waste management, water conservation, reuse of materials… (Rating systems)
Retain historical setting and urban pattern	Extent of urban pattern and character of the area (context of the building) preserved in the case of land use changes
Infrastructure	Extent of adequate infrastructure, transport and amenities provided in the proximity of the area
Townscape	Buildings, spaces and envelopes, the connections and closures, the vistas and views all knit together to form a set of harmonious relationships
Political	
Community participation	Extent of participation of the affected population in the formulation and execution of the adaptive reuse of the historic building
Supportive policies	Extent of Government incentives, regulations, agreement and funding
Transparency and accountability	Extent of the formulation process of the policy known by the public

Source Yung and Chan (2012), revised by the author)

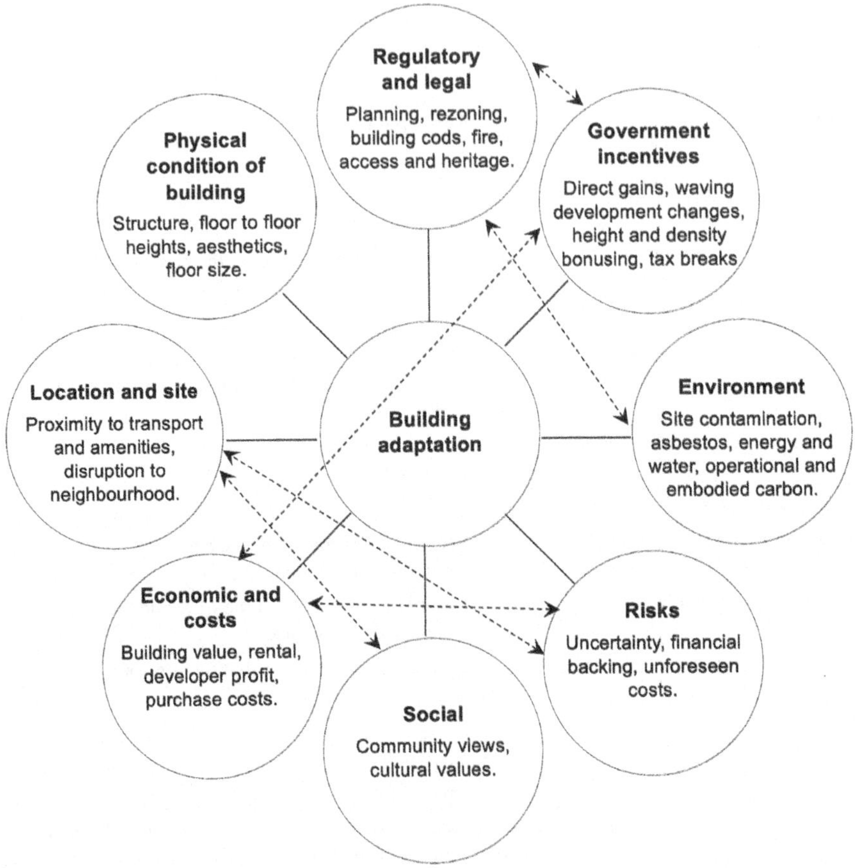

Fig. 2.8 Model of decision-making factors in building adaptation. *Source* Wilkinson et al. (2014), revised by the author

- ***Building design***—Environmental sustainability cannot be achieved if development exceeds the carrying capacity of an urban area. To avoid negative effects on the microclimate, the density, height and mass of buildings should be properly planned.
- ***Provision of open spaces***—Open spaces are undoubtedly important for improving the overall environmental quality in a densely built environment. Open spaces are considered the lungs of urban areas as they provide a break in a dense environment.
- ***Rehabilitation of reparable properties***—Renovation of dilapidated properties can significantly reduce demolition and construction waste. In addition, repairing and maintaining existing structures consumes fewer building materials and natural resources than new buildings.

The set of performance indicators developed and adopted to measure the six criteria is presented in Table 2.4. The indicators play an important role as they are able to assess how the renovation project is achieving its environmental objective.

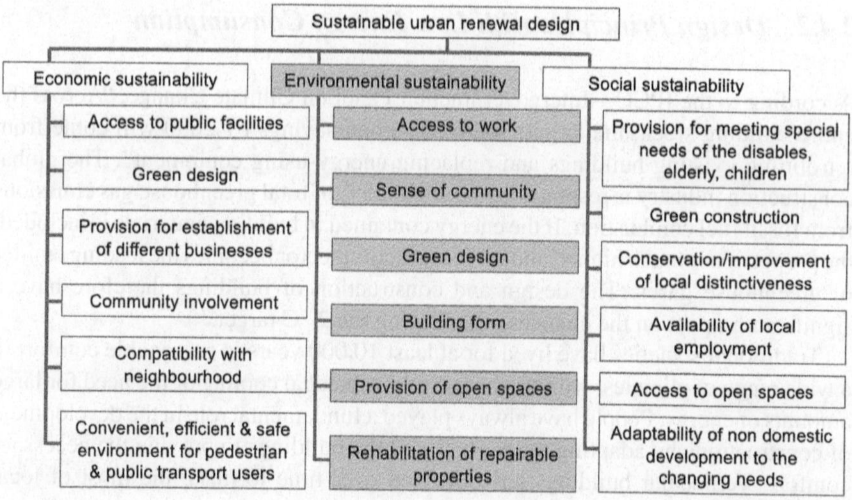

Fig. 2.9 Framework of sustainable urban renewal model. *Source* Chan and Lee (2009), revised by the author

Table 2.4 Twelve indicators derived for six environmentally sustainable design criteria

Performance indicators			
Design criteria indicator	Indicators	Categories	
		Quantitative	*Qualitative*
Access to work	Work travelling habits	X	
	Average time for citizens to get to work	X	
Sense of community	Social cohesion		X
	Citizens' satisfaction with the local community		X
Green design	Incorporation of passive design	X	
	Quality of passive design		X
Building form	Density of development within renewal site	X	
	Quality of building development		X
Provision of open spaces	Percentage of open spaces being provided	X	
	Quality of open spaces		X
Rehabilitation of repairable properties	Percentage of existing properties being retained	X	
	Degree of rehabilitation		X

Source Chan and Lee (2009), revised by the author

2.4.2 Design Principles with Low Energy Consumption

According to the IPCC—Intergovernmental Panel on Climate Change, "Across the entire building stock, the largest share of carbon savings by 2030 will come from retrofitting existing buildings and replacing energy-using equipment". The global construction industry is responsible for about 25% of total greenhouse gas emissions from fossil fuel combustion. If the energy contained in building materials is included, the proportion is much higher, and the construction sector becomes the leading emitter of greenhouse gases. The design and construction of buildings therefore have a significant impact on the chances of achieving the 2 °C target.

Traditional societies have lived for at least 10,000 years in reasonable comfort in a wide range of climates, without the use of mechanical cooling or the need for large amounts of energy. People have always played a fundamental role in the development of constructions by adapting themselves and the buildings to provide the necessary comfort. Vernacular buildings have evolved over time to make the most of local materials and conditions.

As with the design and construction of new buildings, renovation and adaptive reuse must also consider these strategies.

The focus is on buildings that produce a high level of environmental sustainability:

- high thermal comfort (temperature and humidity, heating and cooling),
- visual comfort (light and shade),
- high air quality (fresh air, elimination of pollution, excess heat, etc.),
- architectural quality.

They represent a range of passive strategies involving geometric qualities, envelope and choice of materials, and the use of natural elements such as the sun, light, temperature and air. Passive solar heating or passive cooling uses solar energy to maximise heating or cooling based on a building's exposure to sunlight.

Active strategies include technical systems (lighting, heating, ventilation and cooling) integrated with passive solutions to create a comfortable indoor environment.

A bioclimatic building is fully integrated with natural cycles and able to use them without causing harm. The interaction of the building with elements such as soil, water, sun and air must be carefully considered and integrated into renovation projects.

Passive design strategies are based on climatic considerations and attempt to control comfort without consuming fuel. Typically, the principles of sustainable design apply:

- orientation of the building to maximise the relationship with the sun and natural resources,
- design of the building to take into account its relationship to climate characteristics,
- materials and construction technologies to consider air tightness, heat retention, insulation, and thermal mass,

- use of free solar energy for heating and lighting,
- use of wind and breeze for cooling,
- use of shade to control heat gains.

The issue is how to apply these principles to existing buildings, their renovation and reuse.

Depending on the specific local climate, different seasonal strategies can be used to upgrade existing buildings and transform them into more sustainable buildings:

Strategies for Winter

- Increase the thermal performance of the envelope,
- utilise free solar gains,
- natural ventilation control,
- mechanical ventilation (heat recovery).

Strategies for Summer

- Adaptable transparent surfaces,
- natural ventilation (night, inter-seasonal),
- thermal mass.

Further considerations are discussed in Chap. 3.

2.5 Conclusion

Defining a rigorous process for renovating the built environment is fundamental. The more the building is part of the cultural heritage, the more important it is to carefully evaluate its characteristics and state of conservation, activating the knowledge pathway that allows the right balance to be found between the historical value and innovation during the project phase.

The renovation project must set the same goals of sustainability, environmental protection, and energy conservation that we now apply to the construction of new buildings. In many regions of the world, most buildings are already under construction, often with criteria that do not take into account minimum standards of sustainability. It is therefore the responsibility of the renewal and conversion process of the building stock to close this gap and achieve the 2050 target.

References

Antucheviciene, J., & Zavadskas, E. K. (2008). Modelling multidimensional redevelopment of derelict buildings. *International Journal Environment and Pollution, 35.*

Ball, R. M. (2002). Re-use potential and vacant industrial premises: Revisiting the regeneration issue in stoke-on-trent. *Journal of Property Research, 19.*

Chan, E. H. W., & Lee G. K. L. (2009). Indicators for evaluating environmental performance of the Hong Kong urban renewal projects. *Facilities, 17.*

Ferreira, J., Pinheiro, M. D., & Brito, J. (2013). Refurbishment decision support tools review. Energy and life cycle as key aspects to sustainable refurbishment projects. *Energy Policy, 62.*

Harris, S. Y. (2001). *Building pathology. Deterioration, diagnostics, and intervention.* John Wiley & Sons, Inc.

Nielsen, A. N., Jensen, R. L., Larsen, T. S., & Nissen, S. B. (2016). Early stage decision support for sustainable building renovation—A review. *Building and Environment, 103.*

Thuvander, L., Femenías, P., Mjörnell, K., & Meiling, P. (2012) Unveiling the pro-cess of sustainable renovation. *Journal of Sustainability, 4.* Open access www.mdpi.com/journal/sustainability.

Watt, D. (2007). *Building pathology.* Blackwell Publishing.

Wilkinson, S., Remøy, H., & Langston, C. (2014). *Sustainable building adaptation. Innovation in decision-making.* Wiley Blackwell Editor.

Wilkinson, S. (2012). Analysing sustainable retrofit potential in premium office buildings. *Structural Survey, 30.*

Yung, E. H. K., & Chan, E. H. W. (2012). Implementation challenges to the adaptive reuse of heritage buildings: Towards the goals of sustainable, low carbon cities. *Habitat International, 36.*

Chapter 3
Building Adaptive Reuse

Abstract Adaptive reuse of old buildings has become an increasingly common feature of urban regeneration; reuse of historic buildings should also incorporate energy efficiency, low carbon emissions, improved environmental performance, economic sustainability and cultural identity. Many current buildings will still be around in the next 100 years, so strategies that combine energy efficiency with adaptive reuse of historic buildings must become a codified process. This chapter analyses some good practices and strategies for defining the best reuse solutions.

Keywords Adaptive reuse · Design strategies · Remodelling tactics · Sustainability

3.1 Introduction

Although buildings have always been reused, today's architectural renovations are often seen as a creative way to renew existing historical contexts while reinventing their economic and social value. The combination of new and old architecture must ensure the preservation of an authentic historical character, while a new appropriate use breathes new life into the building. Douglas (2006) defined adaptive reuse as all the construction work and interventions to modify its capacity, function or performance in order to adapt, reuse or improve a building to new conditions or requirements.

Finding the most appropriate function, considering the impact on the context, while taking into account the historical and cultural value, are crucial tasks.

Several aspects need to be considered when deciding on adaptive reuse; successful reuse must respect the existing building and at the same time give the character of the heritage a contemporary level.

The choice of a new use should be based on analytical and scientific methods, taking into account various elements in order to find the most appropriate strategy for the best design of the reuse. Bullen and Love (2011) have analysed the main factors and barriers that influence decisions on reuse, as shown in Fig. 3.1.

M. Grecchi, *Building Renovation*, PoliMI SpringerBriefs,
https://doi.org/10.1007/978-3-030-89836-6_3

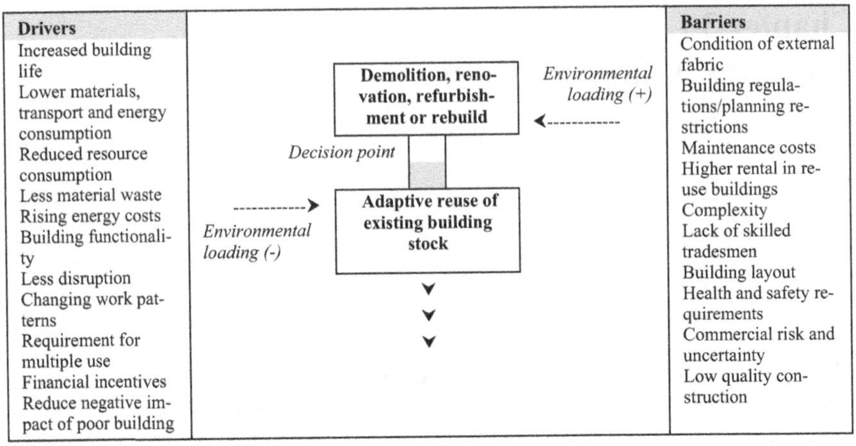

Fig. 3.1 Drivers and barriers of adaptive reuse. *Source* Bullen and Love (2011), revised by the author

Ellison and Sayce (2007) found that buildings that include flexible spaces can be adapted to changing needs, require less frequent and less costly renovation, and remain sustainable over time.

3.2 Factors that Influence Adaptive Reuse

The methodology proposed by Misirlisoy and Günçe (2016) can be adapted and applied to develop reuse strategies in any context, taking into account local policy and planning issues. The model considers three main phases in collecting data at different scales and in knowing local identities, some of which are collected through interviews. This information will form the basis for planning possible renovations that are compatible with the buildings and useful for regenerating a context (Table 3.1).

Table 3.1 groups the complexity of factors influencing decisions in the project phase into five categories. Based on the analysis of the building, which includes its original functions, physical features, heritage history and value, and neighbourhood characteristics, the adaptive reuse potential can be hypothesised. Building height and width, floor size, utility core, engineering grid, and supporting structures are the elements that typically determine whether adaptation is possible. Misirlisoy and Günçe defined four phases with subsequent actions:

1. This can include securing the building without a complete restoration
2. The full restoration of the building, including consolidation and completion of missing parts
3. Proposals for new construction and reconstruction of the building
4. The combination of two or three phases together.

Table 3.1 Factors affecting adaptive reuse

Factors affecting adaptive reuse		
1	Actors	
2	Analysis of existing fabric	
	2.1	Original function of the building
	2.2	Physical character
	2.3	Heritage values
	2.4	Needs of the district
3	Conservation actions	
4	Adaptive reuse potentials	
5	Functional changes	

Source Misirlisoy and Günçe (2016), revised by the author

After the analysis, the potential for adaptive reuse for the new use should be evaluated and a decision made about the new features. The proposed model can be adapted to develop reuse strategies for any listed building in any context, taking into account local planning issues.

Step 1—Define the actors in the decision-making process

The first step is to define the actors involved in the process. The actors considered by Misirlisoy and Günçe are the stakeholders who will contribute to the decision-making process. These actors are grouped into four categories, as shown in Table 3.2.

Step 2—Analysis of the context data

The second step is the collection and analysis of all information related to the context, because it is certain that the location influences the adaptation (Table 3.3).

The dimension of the context to be analysed may change in the light of specific conditions and possibilities. The scope of the analysis relies on the assumed reuse, depending on what part of the territory is involved. It is generally not a linear process;

Table 3.2 Extract from the Misirlisoy and Günçe's model

Step 1—Definition of actors in decision making			
Users	Producers	Investors	Regulators
Original users	Architect	Owner	Planning authorities
Contextual users	Planner	Tenant	Local authorities
	Engineers	Municipality	Municipality
	Restoration experts	Government	
	Specialists	Funding organization	

Table 3.3 Aspects to be analysed at the context scale

Step 2—Aspects to be detected			
Context (region, city, district scale)			
Mobility	Environment	Social/economic	Urban/Value
Road conditions Public transports Cycle and pedestrian paths Public and private parking	Green areas Nature (parks, see, river, hills, etc.) District heating Climate: Temperature, sun, rain, wind, humidity Air quality / pollution	Main activities Development previsions Demographic trend Market needs Land value Social progress and trends Cultural phenomena	Public spaces Density Landmarks
Site			
Mobility	Environment	Energy	Physical
Accessibility Parking areas	Green areas Soil permeability Noise and acoustics	District heating Water table Wind data Geothermal data Microclimate condition	Location Dimensions Soil characteristics Borders Plot size and Site plan Adjacent building and obstacles

Extract from the Misirlisoy and Günçe's model

several hypotheses need to be evaluated and validated through specific studies of the context before new objectives can be set. Urban planning, local traditions and discriminatory needs must be transformed into project inputs.

An existing purpose may, to some extent, affect the potential of a new or changed purpose. The planning zone legally defines what is considered permissible development in a particular area.

Step 3—Analysis of existing buildings

The third step is the analysis of the existing buildings, which includes the identification of their original purpose, physical characteristics, historical and patrimonial value. The analysis of the physical characteristics (available area, number of floors and size, width, volume and shape), the position of the structure (distance between columns or bearing walls), the construction techniques, the characteristics of the envelope (openings, roof, balconies, materials, etc.) and the installations (location of the service core) are essential to find the new most suitable use for the building (Table 3.4).

Location can influence the adaptive reuse potential (e.g., whether or not a building adjoins other buildings on one or more sides, the number and position of entrances, which may promote flexibility).

Table 3.4 Aspects to be analysed at the building scale

Step 3—Aspects to be detected			
Fabric			
Architecture	Physical	Characteristics/Functional	Environment
Formal characteristics (symmetry, geometry, hierarchy, etc.) Façade characteristics Heritage value Architectural value	Construction technologies and materials Structural integrity and foundations Mechanical systems Deteriorations and obsolescence	Dimensions (surface, volume, number of floors, height, etc.) Structure grid Connections (entrances, corridors, stairs, lifts, etc.) Spatial organization, spatial flow Flexibility Service ducts Open area and garden	Orientation Solar access Glazing and shading Natural lighting and ventilation Energy rating and existing performance Envelope characteristics (insulation, thermal recovery, etc.) Atria, courtyards

Identifying the values of cultural heritage is one of the factors that will influence decisions; specific regulation can impose restrictions on reuse and influence technological solutions.

A SWOT analysis, which could be the final result of the first part of the knowledge and data collection, can also be used as a control method to validate all the hypotheses established.

Step 4—Decision on conservation measures
Once the structural condition has been analysed, the proper preservation measures and repair work must be planned. The actual condition of the building will influence decisions for new use. A historic building may exhibit various levels of deterioration, which will determine a range of repair strategies such as maintenance, restoration, consolidation, demolition, and partial reconstruction. (See Chapter 8 for some examples of technical repair systems). Only after this stage can adjustments such as restructuring, additions and architectural decisions be considered (Table 3.5).

Table 3.5 Extract from the Misirlisoy and Günçe's model

Step 4—Decision of conservation actions			
Phase 1	Phase 2	Phase 3	Combination of phases
Preservation	Restoration	Remodelling	Phase 1 and 2
Refurbishment	Renovation	Transformation	Phase 1 and 3
Rehabilitation	Consolidation	Adaptive reuse	Phase 2 and 3
Maintenance	Reconstruction	Extension	Phase 1, 2 and 3

Step 5—Definition of the adaptive reuse potentials
Analysis of a building's adaptive potential is essential to develop appropriate strategies for sustainable reuse. Misirlisoy and Günçe define these potentials as physical, economic, functional, ecological, political, social and cultural potentials (Table 3.6).

Step 6—Decision on functional changes
There are three main options for an adaptive reuse project:

1. The original functions are retained, with some adjustments necessary to adapt the building to new needs. Typically, this approach involves valuable historical buildings where a change of use would result in a significant change.
2. The adaptation involves a mixed use. The original intended use could be maintained, and new functions could be integrated.
3. Adaptation to entirely new uses. Typically, this approach is possible for historic buildings that are not protected heritage. In some cases, adaptive reuse of a valuable heritage asset is only permitted for "removable works" that preserve the original beauty intact (Table 3.7).

At the end of the process, after assessing the economic feasibility, an adaptive reuse project can be developed (Fig. 3.2).

3.3 Design Approaches

When it is time to renovate a historic building, a variety of approaches can be taken. While there are official definitions, different interventions are often required, motivated by specific conditions such as the historic value of the building, the degree of deterioration, the morphology, and the architectural features. In many cases there is an overlap between the different approaches. Some definitions are given below.

Conversion—Changing the use of a building while maintaining the main configuration.

Conservation—Any action that prevents deterioration and extends the life of the building. This typically involves cultural heritage and the aim is to pass on its theoretical and cultural message to future generations. Many philosophies characterise the conservation process, but in brief they are:

- minimum degree of action,
- reversibility of the added parts,
- evidence of interventions that should never be disguised.

Listed buildings can be adapted to new uses to preserve their heritage value and do not necessarily need to be preserved as they are. The typical reuse of listed buildings is associated with cultural and public functions, but the choice is a delicate one that involves the characteristics of the building, the site, the cost, and the future economic capacity to maintain the structure.

Table 3.6 Extract from the Misirlisoy and Günçe's model

Step 5—Definition of adaptive reuse potentials

Physical	Economical	Functional	Environ	Political	Social	Cultural
Originality of architecture character Aesthetics Disability access Human scale	Site access Population density in the location Profits from market demand Market opportunity Financial resources for maintenance cost	Spatial flow Adaptability Space/structure relationship Flexibility of spaces	Site and location Environmental quality of the surrounding Neighbourhood relationships Orientation of the building	Conservation planning requirement Building regulations Urban master plan Land use plan and zooning Ownership	Social meaning for the local community Spirit of the building Public interest to the building	Cultural meaning for the local community Historic significance Authenticity

Table 3.7 Extract from the Misirlisoy and Günçe's model

Step 6—Decision of functional changes		
Adaptation with some use	Adaptation with mixed us	Adaptation with a totally new use
No change in original function	Keeping the original use and adding some sub functions	Changing the original function

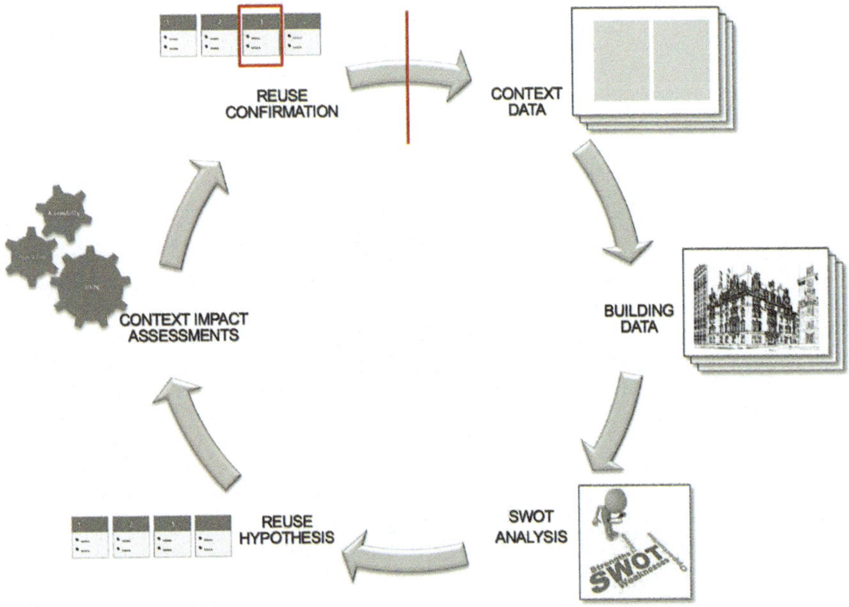

Fig. 3.2 Adaptive reuse process. *Source* The author

Reconstruction—Reconstruction of a building or part of it is the most extreme measure. In the past, there were typical "fakes" of old demolished buildings to restore their appearance and historical memory. A modern approach prefers to reconstruct a new volume, in dialog with the existing volume, but never to reproduce it.

Refurbishment—The term describes a renovation process and the changes required to adapt the building to a new use. It is sometimes seen as an alternative to demolition and reconstruction. Refurbishment is a series of alterations to adapt a building to a new use and improve its performance. Alterations, additions and all repairs must be appropriate to the new use, while preserving the historic and architectural value of the building. Based on the existing features, various methods can be used to convert a building. Every refurbishment begins with an exhaustive analysis and evaluation of many factors and possibilities. A balance must be struck between the memory of the past and future renovation, not only for the building but also for the site in question.

Rehabilitation—A transformation process to allow a change of purpose through renovation, remodelling and additions, while preserving historic and cultural values. This is often seen as the "best way to preserve buildings", to breathe new life into an abandoned building. Rehabilitation is sometimes used as a synonym for the term renovation.

Renovation—Other work is possible if it is not historical and cultural heritage. Typically, a poor structural condition requires upgrading of the building envelope with new cladding, retrofitting, or a new interior design for adaptive reuse, sometimes with significant changes to the supporting structure.

Restoration—Restoration is the process of preserving a historic heritage, sometimes removing recently added parts that have diminished the original value. Antique materials must be respected, and accurate documentation of the building's life span is necessary to identify interventions and decide which should be preserved as "historic additions". Limited rehabilitation is compatible, typically in terms of furnishings and functional requirements. Restoration work should preserve, revitalise, and showcase historic and architectural values. Adaptive reuse, if required, must fully respect the original building.

Retrofit—A retrofit is essentially the insertion of new and more modern systems into an existing building. The term is usually associated with installations and performance. For example, an energy retrofit refers to improving the building performance by introducing layers of insulation, more efficient heating and cooling systems, and the use of renewable energy sources.

3.4 Strategies for Sustainable Building Renovation

Existing buildings, whether part of the cultural heritage or not, can provide an opportunity to transform the intended use of an area and initiate its regeneration. The potential of existing buildings can be considered as part of a sustainable development process, not only from an energy point of view, but also from an economic, social and cultural point of view.

Many examples show how urban regeneration began with the reuse of abandoned buildings, such as the post-industrial revitalisation of docklands, disused railway stations and factories. However, it is difficult to start a regeneration process with the reuse of a single building, but the restoration of a cultural heritage can become a reference point and help to change the identity of a place.

There are many different degrees of alteration, from minor work necessitated by new regulations to a complete redesign with retention of the original façades and new volumes behind. Sometimes remodelling and adapting to new uses is the only way for an old building to survive.

Depending on the new functions and repurposing of the buildings, structural changes may result in minor work or require major structural changes, often even demolition. In addition, an adaptation may completely change the performance of the building or simply improve it, depending on the architectural design goals. Two types of alterations can be considered: those to the building envelope and those to the internal layout. As outlined by Kincaid (2002), four adaptation strategies result from the combination of external and internal changes.

The options for physical changes can be analysed in the following five steps:

- consider different options, from low to high changes,
- eliminate those options that are not technically feasible or economically affordable,
- eliminate those changes that are not the best solution for new destinations,
- restrict the range of options on the basis of specific detailed considerations,
- analyse comparisons between different technical design assessments and select the most affordable scheme for adaptation.

A "low change" strategy could be considered if the characteristics of the building (e.g., size, morphology, structural grid, envelope and height from floor to floor) are compatible with the proposed new uses. In contrast, "high remodelling" is selected when specific investigations and evaluations (e.g., morphology, size, and also structural damage) result in remodelling and improvement of the original structure, typically with partial demolitions and horizontal or vertical extensions.

Selective demolitions involve partial demolitions to allow for adaptation to new uses. Less extensive demolitions will be required for building repairs and renovations, such as walls, systems, cladding, and windows.

To renovate a building is to add value that does not replace the original building, but creates an entity in which the sum of the old and the new defines a better image. When renovating a building, different approaches can be taken by integrating new parts or layers that can be summarised as two- or three-dimensional transformations. Reversibility is often a strategy to transform historic buildings while preserving their value; of course, this is more of a conceptual approach than a real prospect of returning the building to its previous state.

When a building is reused, the most important and significant factor in architectural design is the relationship between old and new. Brooker and Stone (2004) developed three categories of building reuse based on the degree of integration between the existing building and the new elements. These strategies are installation, insertion and intervention, as defined below:

- ***Installation***—The old and the new building exist independently of each other. The new elements are located within the boundaries of the existing building. This is an application of the concept of "reversibility".
- ***Insertion***—A new, independent volume that is precisely adapted to the existing envelope and is located within the boundaries of the building.
- ***Intervention*** (the most symbiotic relationship) – Comprehensive transformations involve the original construction, with additions and changes.

Each building represents a "starting point" from which interventions can be planned that cannot be separated from the original architectural character. A large number of buildings, most of which were built in the last fifty years, show signs of obsolescence. The deficit of the built environment can be briefly expressed as another kind of obsolescence:

- ***Figuratively***—Spatial and pictorial of poor quality.
- ***Functional***—Abandoned buildings that have exhausted their life cycle for functional or technological reasons.
- ***Technological***—Buildings that do not meet the required level of performance.

To renew a building means to add value. The project must start from the building characteristics and find the right renovation strategies, perhaps in architectural contrast rather than continuity. Three main renovation procedures can be considered.

Transformations

Changes affect the whole building; a clear boundary between old and new is not apparent. Many approaches can be shown; it makes sense to focus on concepts related to sustainability and energy saving.

New envelope—New layers can be added to an existing façade to solve energy problems, improve the building performance and change the appearance. Only interior solutions are possible for listed buildings.

In energy retrofits, it is important to design a new building envelope that not only improves the building performance, but also takes into account the various requirements that arise from the new interior layout to meet occupant needs and enhance architectural value. A typical example is the redevelopment of many social housing neighbourhoods that require total refurbishment, as will be seen in Chap. 6 (Fig. 3.3).

Passive design strategies—It is necessary to return to the origins and reintroduce strategies that reduce energy requirements and improve the quality and well-being of the interior of buildings, taking into account the climate and local conditions. These strategies are typically associated with insulation and air tightness, solar radiation, glass performance, daylight and shading systems, and natural ventilation (Figs. 3.4, 3.5, 3.6, 3.7 and 3.8).

Atrium—Atriums have potential to reduce energy consumption, especially when renovating existing buildings.

An atrium can be covered with a lightweight roof to become a thermal buffer zone.

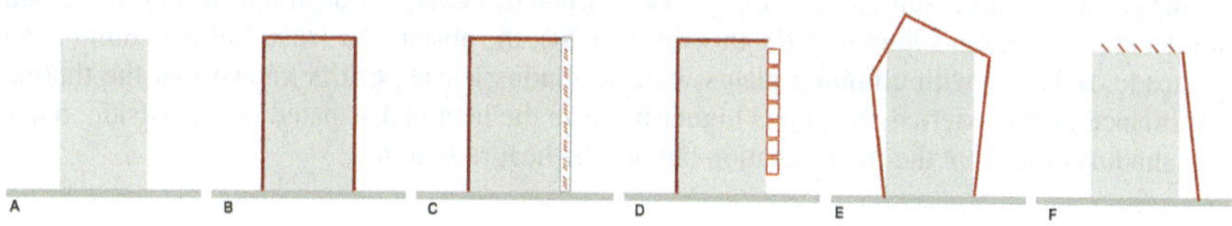

Fig. 3.3 Remodelling tactics. **a** Existing building. **b** New layers increase the performance of the existing envelope. **c** Different solutions depending on the orientation. For example, north-facing insulation combined with a south-facing double skin integrating shading and seasonal natural ventilation. **d** Small greenhouses harness solar energy and create outdoor spaces. **e** Architectural enhancement. **f** Integrated building envelope with renewable energy systems

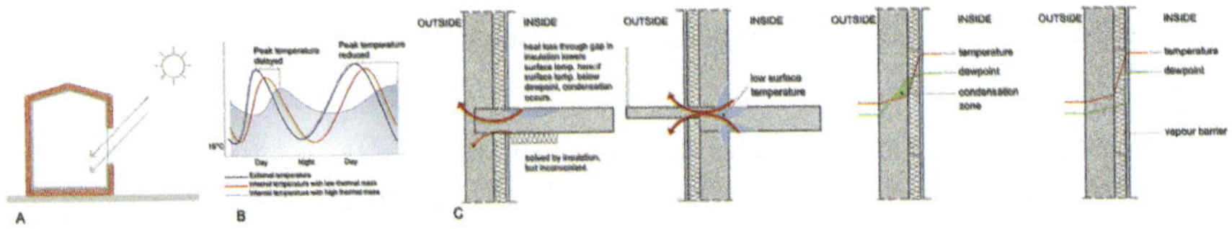

Fig. 3.4 External insulation is preferable as it facilitates the prevention of thermal bridging and surface condensation in a poorly ventilated building. **a** The external insulation protects the envelope from solar gains. The thermal mass remains coupled inside (gains from internal heating and residents). **b** Thermal inertia. Mass is especially beneficial when there is a large difference between daytime and night-time outdoor temperatures. **c** Thermal bridges, surface condensation, thermal mass insulated from the inside, and interstitial condensation that reduces the effectiveness of the insulation are the effects of interior insulation

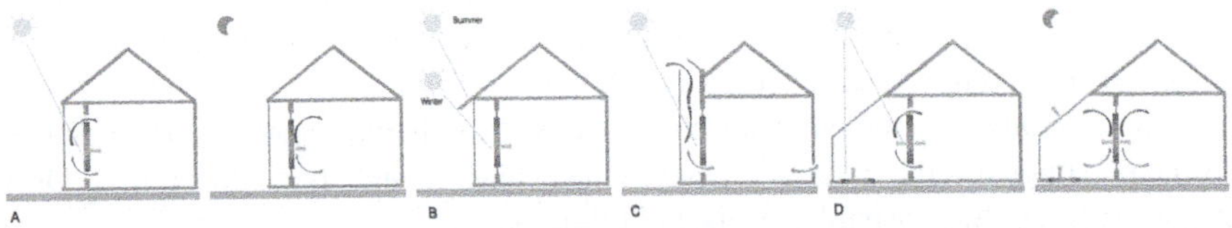

Fig. 3.5 Passive solar heating. All options must be well evaluated depending on the specific climate, as they are not an effective solution under certain conditions. **a** Thermal mass wall (indirect gain). Thermal mass is placed between windows and rooms to absorb heat and allow slow transfer to rooms (Trombe wall) with a convective radiation path. Air outlets can be closed at night to keep convection in the right direction. **b** A Trombe wall with an overhang to provide shade from the summer sun. **c** A Trombe wall that acts as a solar chimney. **d** A sunroom with air intakes for convection and radiant heating (left). The same room at night, with air intakes closed to keep convection going in the right direction (right). Greenhouses are also used as buffer zones

Fig. 3.6 Natural light must be well distributed to avoid discomfort. Areas of high brightness next to areas of low brightness cause glare. Shading devices can significantly reduce building heat gains while maintaining daylight, visibility, and natural ventilation. Conversely, carefully designed shading can allow direct sunlight in during seasons when this energy is desirable. **a** Natural shading with deciduous trees in summer (left). In winter (right), the absence of leaves allows sun to reach the façade. **b** Trellis with climbing plants with deciduous leaves. **c** It is known that the thermal performance of the external shading is higher because the heat is dissipated to the outside. **d** The inner shadings intercept the light radiation but not the heat radiation

A very deep and dark building can be opened in the central part to bring natural light and seasonal ventilation inside.

The contribution of an atrium depends very much on its size, orientation and position. In winter it can reuse and redistribute heat generated in the building, and

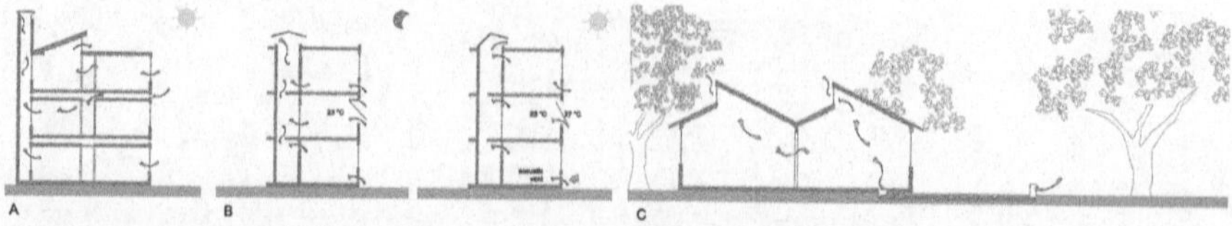

Fig. 3.7 Natural ventilation can be a strategy used at a particular time of year. Once a highly insulated and airtight envelope is created, mechanical ventilation with heat recovery must be used. **a** Chimney ventilation is a passive cooling strategy that takes advantage of temperature stratification. **b** Night ventilation strategy. It works only in the presence of internal thermal mass and high ventilation rates during the night. **c** Ground cooling pipes. Air is cooled (and possibly dehumidified) as it travels. The cooling effect depends on the presence of a sufficient temperature difference between the outside air and the ground

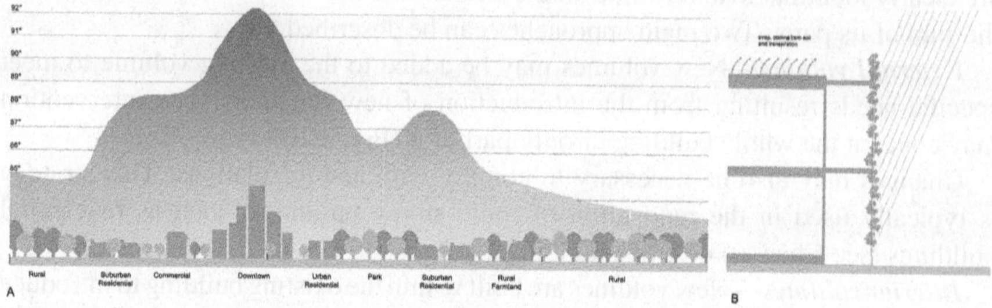

Fig. 3.8 **a** Urban heat island. **b** Use of vegetation to protect the building. In the case of the roof, turf (soil and vegetation) is not a good thermal insulator. The advantage is evaporative cooling. A green façade provides shade and protection in summer

in summer it can distribute heat loads. Two parameters, shelter and solar energy, can be combined to obtain very different performances.

Advantages in winter are:

- natural light in the middle of the volume,
- a protected room with a different temperature than the outside temperature,
- increased solar gains.

The benefits in summer are real when:

- adequate natural ventilation is activated to enhance the adhesive effect,
- glass exposed to sunlight is shaded by mobile equipment.

An atrium can be used for temporary activities, can become a winter garden and can combine permanent activities located around the atrium (Fig. 3.9).

Additional

Many strategies can be used to increase the total volume, with a wide range of design options: Expansion, Enlargement, Extension, Integration and Inclusion. The main goal is to add space, improve the building, or change its appearance, such as when a

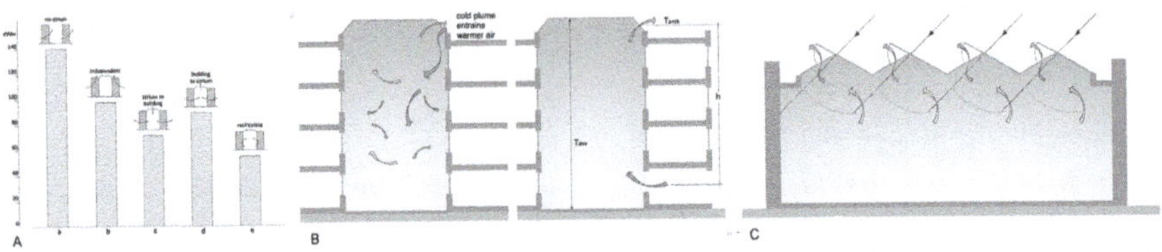

Fig. 3.9 Atrium. **a** Comparison between different atrium models. **b** A high degree of winter ventilation allows mixing of cold air down to the ground floor (left). Upper and lower vents provide maximum ventilation in summer. Layering ensures that the atrium is coldest near the ground floor (right). **c** Use of skylights in a typical case of adaptive reuse of industrial heritage

new skin is added. Successful projects are typically those where the old and the new are clearly identifiable and form a single whole. The new whole must be more than the sum of its parts. Two main approaches can be described.

External volume—New volumes may be added to the existing volume to meet specific needs resulting from the introduction of new functions. The intervention may concern the whole building or only part of it (Fig. 3.10).

Changes may also be necessary to comply with new regulations. This strategy is typically used in the renovation of multi-storey buildings, such as residential buildings (see Chap. 6) or office buildings.

Interior volumes—New volumes are built within the existing building to introduce new functions while maintaining the existing architectural value.

This approach is often applied to cultural or industrial heritage with different objectives. In the first case, the objective is to preserve the original value by adding new, reusable parts; in the second case, the solution also takes into account the size of the building and allows the introduction of new, independent parts that also reduce energy requirements for heating and cooling (see Chap. 4) (Fig. 3.11).

Subtraction

Demolition may also be considered. The reasons why parts of a building are removed can vary widely, sometimes closely related to urban scale, or partial demolition may be necessary because of new functions.

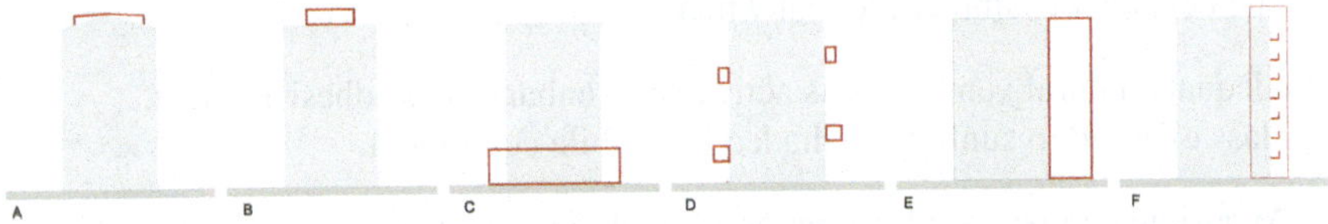

Fig. 3.10 **a** The shape of the roof can be changed to convert an attic into living space. **b** New volumes can be added above a flat roof. **c** The first and ground floors can be converted for new functions that require more space (typically retail and office space). **d** Multiple volumes can be added to the façade to increase interior space. **e** A new block connected to the existing one. **f** Starting with the renovation of the vertical and horizontal distribution, new volumes (elevator and fire escape) and external horizontal connections can integrate and update the old building

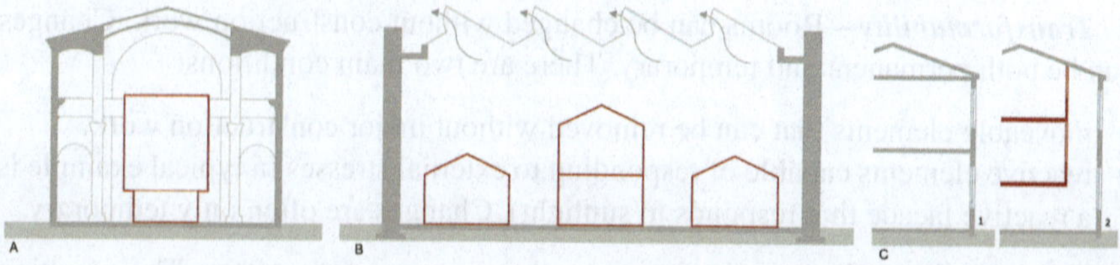

Fig. 3.11 Interior volume. **a** The figure represents a solution chosen for a listed historical building. **b** Typical example of inserted volumes in a former industrial building. The existing envelope provides protection for new heated and cooled volumes. **c** In historic multi-storey buildings it is often possible to divide the space into several levels. The solution must take into account the original façade, which, if it is a listed building, must not be altered

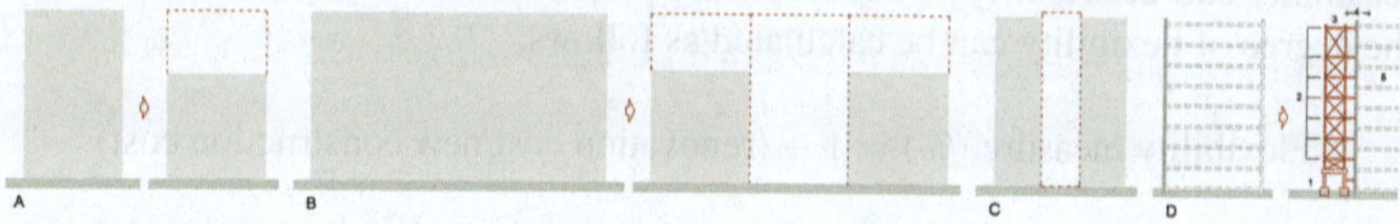

Fig. 3.12 Several possible volume reductions. **a** Partial demolition, e.g., to reduce the number of storeys. **b** Drastic volume reduction, typically to create a better urban balance (see S. Forster's example in Chapter 6). **c** Interior demolition, for example to create a sheltered courtyard or atrium. **d** 'Façadism' is typically considered when it is necessary to preserve the historic value of the facade and the existing building cannot be reused (Construction phases: 1. concrete blocks as temporary foundations. 2. lateral bracing. 3. retaining structure of the façade. 4. system to compensate for rotations caused by wind load, differential failure or other factors. 5. demolition of the volume behind the façade)

The extreme strategy is called 'façadism', which means that only some external facades, e.g., facing urban streets, are retained, usually because of their historic value or the opportunities offered by local regulations (Fig. 3.12).

3.5 The Need for Flexibility

One of the most useful design approaches to adaptive reuse is flexibility; its integration into the renovation project allows for changes to meet new needs in the future. Flexibility can be considered active (e.g., movable walls), but can also include inherently flexible features such as multipurpose spaces, great rooms' floor-to-ceiling height that can be divided in the future, and voids for implementing and repositioning systems.

Depending on the extent of change and the degree of permanence, flexibility can mainly involve three levels.

Adaptability—The ability of a room to support multiple functions without changing the building. Examples of this are repositionable furniture and multifunctional rooms. Changes are not permanent.

Transformability—Rooms can be changed without construction work. Changes can be both permanent and temporary. There are two main conditions:

- moveable elements that can be removed without major construction work,
- reactive elements capable of responding to external stresses (a typical example is a reactive façade that responds to sunlight). Changes are often only temporary.

Convertibility—Changes in functions that require some effort. The resulting changes are often permanent. When planning alterations, it may be useful to consider possible degrees of flexibility. For example, this could include additional space for possible future needs, the basis for future expansion, and investment in additional facilities.

Determining how flexible or adaptable an existing building is could help assess the feasibility and desirability of a remodelling project. According to Blakstad (2001), the degree of flexibility can be calculated as follows:

$$\text{Flexibility measure (\%)} = 1 - (\text{renovation cost/new construction cost})$$

Two main types of flexibility can be considered.

Flexibility of use—This refers to how well the space can be adapted to the changing needs of the users. Changes can be requested at different intervals (daily or after a longer period of time). Simple devices can change the space quickly (typically sliding walls), but there are also structural and morphological conditions that promote flexibility (Fig. 3.13).

Flexibility over time—Making initial arrangements for possible future adaptations. Adaptive reuse will be easier if flexibility is considered in the original project. Analysing the characteristics of existing spaces and the conditions that favour flexibility is essential to understanding the propensity for change in a building (Fig. 3.14).

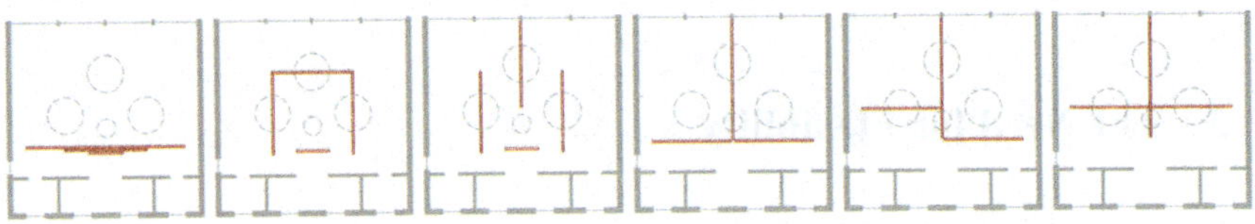

Fig. 3.13 Operational versatility. Typically adopted in residential units or offices

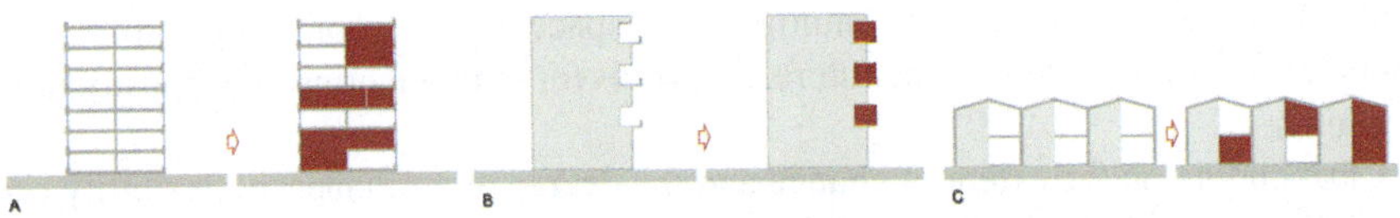

Fig. 3.14 Flexibility over time. **a** All modules can be easily reorganised, replaced, combined or exchanged. **b** Enlargement of the usable area, whereby the total volume of the building remains unchanged. **c** Enlargement of the volume. The initial arrangement allows easy implementation

3.6 Conclusion

Finding a new use, considering the impact on the context and the necessary transformations, while respecting the historical and cultural value, are complex tasks.

The perspective of reusing existing buildings can be seen as part of a sustainable urban regeneration process, not only with an energy objective, but also from an economic, social and cultural point of view.

The definition of new uses must be based on analytical and scientific methods, taking into account factors at different levels in order to develop the most appropriate strategy and project for reuse.

When it comes to renovating a historic building, different approaches can be taken. Official definitions can be found in the literature, but often interventions require the intersection of different works and principles. The key is the rigour with which decisions are made to ensure the preservation of the original historic and architectural values that are to be retained and enhanced. The combination of new and old architecture must preserve the original historic character, while a new appropriate use breathes new life into the building.

There are many options for renovating an existing building, depending on several factors, such as quality and historic value, building characteristics, relationship to context, and new proposed use. These options have been grouped into three main tactics; the first is called "Transformation", which involves two- or three-dimensional interventions in the building envelope. Its priority is energy retrofit. The second, called "Addition", groups all the actions related to the increase of the volume. Finally, "Subtraction" is the third approach, where certain parts are demolished, for example to reduce urban density or to create internal voids to improve comfort.

One of the most useful planning approaches for adaptive reuse is certainly flexibility; its integration into the renovation project allows for new transformation and use in the future.

References

Blakstad, H. S. (2001). A strategic approach to adaptability in office buildings, Doctoral Thesis, Norwegian University of Science and Technology, URN: NBN, No-2306.

Brooker G., & Stone S. (2004). *Re-readings. Interior architecture and the design principles of remodelling existing buildings*. RIBA Enterprises.

Douglas, J. (2006). *Building adaptation* (2nd ed.). Spon Press.

Ellison, L., & Sayce, S. (2007). Establishing sustainability criteria relevant for the commercial property investment sector. *Journal of Property Management, 25*.

Kincaid, D. (2002). *Adapting buildings for changing uses*. Spon Press.

Misirlisoy, D., & Günçe, K. (2016). Adaptive reuse strategies for heritage buildings: A holistic approach. *Sustainable Cities and Society, 26*.

Chapter 4
Industrial Heritage: Sustainable Adaptive Reuse

Abstract The adaptive reuse of abandoned industrial heritage plays a strategic role that not only affects buildings but can also stimulate a renewal process in large urban areas. The typical decay of many suburbs, which experienced a great development in the period of the Industrial Revolution, has become a predominantly social problem. Different types of reuse have different impacts on industrial heritage. The new uses and the scale of change required must be compatible with and appropriate to the historic significance of the site. The renovation of former industrial buildings, while respecting and preserving the cultural heritage, must not ignore the aspects of energy saving and comfort that are necessary today. The examples presented in this chapter illustrate some interesting renovation approaches that also address complex challenges from a sustainability perspective.

Keywords Industrial heritage · Urban regeneration · Adaptive reuse · Energy saving and performance

4.1 Introduction

Abandoned, run-down and peripheral areas are often more of a social problem than a building problem. Major changes, for example in the production process, have led to major changes in urban planning, opening up new opportunities for planners to think about models and mixes of functions, and to understand the mistakes of the past.

Urban regeneration, with different solutions to the specificities of places, is therefore a challenge that can create new opportunities for communities, resolve critical issues and test new centralities within the city that are better suited to new jobs and the needs of city dwellers. The renovation of disused industrial areas offers the opportunity to redevelop and transform parts of the city into new functions capable of attracting and virtuously activating a process of renewal.

The subject is certainly complex; it is not obvious what mix of functions can be successful, especially as it is necessary to act at different levels, taking into account the urban and social context. The examples are numerous, but often do not show a real success that meets the expectations of transformation. It is often the charm of the

M. Grecchi, *Building Renovation*, PoliMI SpringerBriefs,
https://doi.org/10.1007/978-3-030-89836-6_4

industrial heritage rather than the context that guides some decisions. Enlightened operators or, on the contrary, spontaneous transformations have shown the greatness and potential of these marginal areas. A balance between historical memory and innovation can be the recipe for success.

Since the 1980s, as a result of new processes involving profound changes in the productive sector, primary industrial activities such as metallurgy, chemistry, mining, shipyards and heavy industry have been gradually abandoned, especially in countries where their Industrial Revolution generated enormous expansion and immigration phenomena. This process made available large areas of land, often in neighbourhoods no longer on the periphery, which acquired a high strategic value. Some of them became part of the cultural heritage with restrictions on possible interventions.

Given the obvious economic and social consequences that sustainable urban development inevitably entails, revitalisation is now an opportunity of great interest. Good planning and the creation of specific processes are the basis for a successful outcome, an opportunity for a regenerative model that pays attention to environmental quality and sustainability and introduces an effective mix of new activities, functions and users.

There are many opportunities for industrial buildings to be repurposed. Some of the most popular conversions include museums, art studios, work units, offices, apartments, schools, shopping malls, and, increasingly, multifunctional mixed-use developments.

This chapter explores which adaptive conversions encourage the revitalisation of disused industrial sites, thereby initiating a wider process of regeneration and upgrading of run-down neighbourhoods with social and economic problems. An appropriate balance between new functions and the valorisation of industrial historical memory must guide decisions—a process of renovating industrial buildings with a change of use, keeping the original design visible as much as possible while improving performance to meet current standards.

4.2 Cultural and Social Value of Abandoned Industrial Buildings

Terms such as 'industrial heritage' first appeared in the heritage debate in the 1970s with the establishment of North American, British and international industrial archaeological associations (Falconer, 2005). However, it was not until the beginning of this century that the first global and widespread methodological approaches were developed, and it was not until 2011 that the first international standards for the conservation and reuse of industrial buildings were established.

Adaptive reuse, then, was a practical rather than a theoretical process in the twentieth century, one in which communities in the United States and Europe took advantage of the cheap availability of large unused spaces for underground or marginal activities, sometimes simply by occupying abandoned factories. Industrial buildings and their outdoor spaces represent a unique and sometimes neglected part of our cultural heritage. Since the 1960s, the old abandoned factories have become an opportunity to develop important transformations of the territory. Thanks to their typical size, they represent a value for the cities marked by the industrial development of the nineteenth and twentieth centuries. The preservation of these icons is an opportunity to witness an important historical period and at the same time regenerate places that are usually very run down. The success of the regeneration of former industrial areas depends very much on the degree to which they are integrated into the city and their ability to create cultural, economic and social benefits.

Banham (1986) was the first to emphasise the architectural character of industrial buildings, derived entirely from technical solutions and the evolution of building since they first appeared in the mid-eighteenth century Northern Europe. England was the first and most advanced country in the use of iron in construction from the mid-eighteenth to the mid-nineteenth century, while France had achieved an equal position in the use of concrete. Throughout the nineteenth century, industrial buildings were the main typology for architectural experimentation. Structural innovations were tested on this type of building, and it was on them that a new architectural canon was defined. One of their most recognisable formal features is the organisation of interior spaces in the form of a hall, achieved thanks to slender steel columns and then in concrete. Other formal features are the numerous large windows, unprecedented in their arrangement and size compared to other contemporary building types.

In its "Recommendations on the protection and conservation of industrial, technological and Civil Engineering Heritage" (1990), the European Council defined measures for the protection of old industrial buildings in accordance with their role in post-industrial society. "The aim is to take into account not only buildings, technical monuments, sites or objects, but also a physical environment, a corpus of knowledge, techniques and ways of life". The recommendations provide guidance on four main issues:

- The first concerns the need to identify, study and analyse this type of heritage systematically, using "modern and efficient methods and resources according to common criteria defined at European level".
- The second concerns the protection and conservation of this heritage. Here the emphasis is on the need for selection, which "must be made in such a way as to ensure a balanced representation of the various branches of production". The recommendations also encourage strategies to create incentives.
- The third measure consists of "making the public aware of the technical, industrial and building heritage", promoting training on subjects related to this heritage at all levels.
- The last measure is to encourage cooperation and financial support for possible local projects of European significance.

According to Stratton (2000), the United Kingdom was one of the first nations to experience industrial decline in traditional manufacturing and dock management. As a result, the United Kingdom pioneered the preservation and reuse of redundant warehouses and factories. "The industrial heritage consists of the remains of industrial culture that are of historical, technological, social, architectural or scientific value. These remains consist of buildings and machinery, workshops, mills and factories, mines and sites for processing and refining, warehouses and stores, places where energy is produced, transmitted and used, transport and all its infrastructure, and places used for social activities related to industry such as housing, religious worship or education. Industrial heritage can play an important role in the economic regeneration of decayed or declining areas. The continuity that reuse brings can provide psychological stability to communities facing the sudden end of longstanding sources of employment".

Adaptive reuse of industrial buildings includes projects by public actors whose primary interest is also social, as well as by private entrepreneurs seeking their own profit. Apart from cases where economic and practical reasons have been the main drivers for conservation and reuse, the character of industrial buildings is often interpreted as a different and attractive environment for art and cultural activities, in the way buildings capture both the past and the present. Industrial buildings are part of our cultural heritage and represent the history of our society. Therefore, in most cases, they play an important role in defining the identity of the place and its inhabitants. The reuse of these buildings has become a trend in many countries around the world in the last decade. There is a tendency to see this as a new phenomenon. However, Stratton (2000) emphasises that the reuse and adaptation of industrial buildings is as old as the Industrial Revolution. The main difference between past and present adaptive reuse is the issue of protection and preservation and urban renewal, which often enters into the decision-making process.

Areas of industrial heritage may have been abandoned long ago, have undergone many changes of use over the years, or have only recently ceased to be used for their original purpose.

Industrial heritage concerns historic buildings whose value could be:

- historical evidence of the Industrial Revolution,
- the dimensions of the building, typical of heavy industry, with empty and huge volumes, necessary for machinery and the production cycle,
- important industrial activities that caused the development of the site,
- architectural reference value,
- technical building equipment that highlights the structural and technological development.

The best way to preserve a building, structure, or place is to use it. Adaptation connects the past to the present and projects into the future. It is important to remember that adaptive reuse is not limited to individual buildings or small areas. Large urban areas can also be subject to adaptive reuse.

Transforming old, vacant factories requires a variety of analyses, from structural features to urban design scenarios. The first examples from the past, typical of cities like New York and London, are spontaneous transformations by young people and artists who turned the old red brick factories of the nineteenth century into studios and lofts where they could work and live. In the past, huge industrial areas were completely demolished and redeveloped with new functions, such as the docks in many important European ports.

Industrial sites are often brownfields, which means that their reuse is hampered by pollution, so sustainability aspects need to be carefully considered when renovating buildings.

Supporting structures are often the most interesting elements for conservation. It is important to identify their materials and deterioration before considering a possible renovation project. Metals were widely used as structures in nineteenth century manufacturing buildings and can be classified into three main alloys based on carbon content: iron, steel, and cast iron, whose structural behaviour varies. Cast iron resists compressive forces very well but is also very brittle due to its high carbon content. For this reason, it has been used mainly for columns and compressed elements of composite iron beams. The structural use of iron became prevalent in the mid-nineteenth century when railroad girders, channels, and I-beams became commercially available. Beginning in the early 1900s, steel structures became the most commonly used structural system.

In reusing industrial heritage, the new project should also aim to preserve evidence of technologies, work processes and flows of materials and people. Adaptive reuse of industrial sites presents a variety of opportunities and challenges. Some of these are similar to those addressed in the renovation of other heritage buildings and sites; others are more specific to the industrial context.

These sites play an important role in community life; they provide tangible links to the past and may have contributed to the livelihood of a significant proportion of the community. Communities and individuals attach different values to industrial heritage; for some they are a source of pride and important memories, for others these buildings are an unattractive reminder of decadence and decay. The choice of design depends on many factors, including the characteristics of the building, the social and political context, the recent history of the site and the need for development.

A very well-known international example is New York's High Line, the conversion of an old freight railroad into a linear park that is now very popular. This project was initiated by two local residents who successfully applied pressure to preserve the site and then worked with the City of New York to develop the park. The "High-Line Friends" continue to maintain and manage the park.

Industrial heritage can play an important role in urban regeneration by reinforcing the character and identity of a place. Abandoned factories can have a depressing effect on surrounding areas and, in large inaccessible areas, act as cracks in the urban fabric, exacerbating social and safety problems.

4.3 Adaptive Reuse as a Driver for Urban Regeneration

There are three main questions that need to be considered in order to assess the appropriateness of adaptive reuse of industrial heritage in relation to the wider urban context:

- How will the reuse contribute to the understanding of the wider urban environment?
- How will the reuse complement and/or contribute to the development of the context?
- If adjacent areas are still in use for industrial purposes, will the proposed conversion detract from or support these?

Broader strategies and approaches to reuse whole peripheries are also important for regeneration. This is large-scale adaptive reuse. At the international level, one of the most significant examples is the Ruhr region in Germany, where a large regional Industrial Heritage Trail has been transformed from an economy based on the steel and coal industries to a new economic system.

Different types of conversion have different impacts on industrial heritage. The new uses and the extent of change required must be compatible with and appropriate to the historic significance of the site.

Environmentally sensitive design can also bring significant additional regeneration benefits. In order to gain the necessary knowledge to properly assess the options for change of use and intervention, a list of further questions can form the basis for downstream decisions. These questions can be arranged by specific topics.

Environment

- How does the contamination of the site affect the potential for reuse?
- What structural and architectural condition of the building is desirable to comply with building codes?

Location/Real Estate Market

- How can the location and accessibility of the building complex affect the adaptive reuse decision?
- How important is the analysis of the real estate value of the site for the selection of a possible conversion?

Regulations

- What influence do historical constraints have on reuse strategy?
- What influence do regulations and urban design guidelines have on decisions?

Financial

- What is the expected economic commitment?
- Are there financial incentives?
- What is the expected economic return in terms of potential reuse?

According to Vardopoulos (2019), adaptive reuse of industrial heritage, if the right strategies are applied, can achieve "an optimal balance between initial investment, energy savings, minimising environmental impact, preservation of heritage and urban renewal".

It is not always possible to find long-term reuse that is financially sustainable and commensurate with the heritage value. In such cases, phased development, interim destinations or temporary closure may be appropriate solutions until a suitable new use is found. Temporary uses can be a way of keeping sites active and avoiding their decline, as all derelict buildings experience accelerated decay (Table 4.1).

In summary, there are four main aspects to consider when discussing adaptive reuse:

Cultural heritage as a source of value—The term 'cultural heritage' is a key phrase when talking about cultural heritage as a catalyst for regeneration. The use of cultural heritage within a place can, in most cases, enrich the experience of the surrounding area and make people proud of the place in which they live. Using the historic environment as a catalyst for regeneration and development will undoubtedly give people a stronger sense of belonging.

Combining change of use and conservation—There are a number of stakeholders involved in any regeneration project. These stakeholders have different needs and desires for new uses of former industrial buildings. For this reason, finding sustainable uses is a complex and in many cases difficult task, as is the conversion process itself. When deciding on the conversion of an industrial building, it is necessary to define how much of the original exterior and interior needs to be preserved or restored and how much needs to be renovated. According to Kincaid (2002), when looking for sustainable uses for redundant buildings, some aspects need to be evaluated and investigated before starting the conversion:

- *Characteristics*—Building location, facilities and support services, physical capabilities and limitations.
- *Use requirements*—The set of use requirements by function and type of use that describes the user's needs based on demand.
- *Performance requirements*—The link between the existing building and the need that balances operational requirements with physical conditions.
- *Decision making process*—Alternative options for change can be evaluated based on economic, physical, and use feasibility.

Interpretation of buildings and industrial sites—Industrial buildings represent a large part of our history. In many communities, factories are an important part of their identity and they act as important local landmarks. Therefore, interpretive considerations must be taken into account when adapting an industrial building for reuse.

Table 4.1 Factors affecting local sustainable development via adaptive reuse projects

Dimension	Factor		Description
Economy	F1	Economic growth	• Growing investments • Local business and market creation • Creating jobs • Taxpaying • Increasing property values • Relieved demand on local authorities
	F2	Tourism (cultural) economic growth	• Attractive cities • Visitors' sentiment and architecture • Promoting cultural tourism
	F3	Local value enhancement	• Through their variety, character and a sense of familiarity and safety
Society	F4	Quality of life improvement	• Income growth • Environmental quality • Healthy and hospitable environment • Health • Safety • Leisure • Resilient and sustainable city of residence
	F5	Community action and involvement empowerment	• Cultural sites provide programming that examine homelessness, poverty, and the need for social networks to support community and inspire people to personal and collective action
Environment	F6	Environmental management	• Climate change mitigation • Eco-building • Energy efficiency • Renewable energy systems • Extending life cycle for buildings, materials and resources • Landfill demolition waste reduction • Reduce GHGs • Reduce resources consumption • Recycling
	F7	Land conservation against urban sprawl	• Reduce land use • Prevent urban sprawl • Reduce suburban sprawl

(continued)

Table 4.1 (continued)

Dimension	Factor		Description
Culture	F8	Public environmental awareness and education	• Contribution to educational potential, cultural skills and knowledge • Public environmental awareness
	F9	Technological innovation	• Technological innovation integration • Recovering local traditional construction techniques
	F10	Local memory and cultural identity preservation	• Maintain local memory • Identity • Diversity • Vitality • Aesthetic enhancement • Retention of visual amenity
	F11	Cultural heritage protection	• Safeguard the words' cultural and natural heritage • Heritage preservation • Industrial past

Source Vardopoulos (2019), revised by the author

Involving the local community—Regeneration projects based on cultural heritage have a better chance of success if residents are involved in discussions and decision-making. Scepticism and negativity towards a development and regeneration project is likely to manifest itself in the local community if decisions are made over the heads of local residents. Landorf (2009) highlights two key principles for sustainable development: long-term planning and multi-stakeholder involvement in the process. Resident initiatives and engagement are critical to making development socially sustainable, and community participation in the adaptive reuse decision-making process is particularly important (Oevermann et al., 2016). However, for obvious reasons, it is not possible to involve the local community in all decisions, but it is strategically important to do so in relation to certain aspects that directly affect them.

4.4 Successful Adaptive Reuse Process

When exploring new uses, it is important to examine how they can be accommodated in existing spaces. Questions to be considered include:

- Will the original floor plan still be legible in the adapted building? Or will reuse require significant changes to the main rooms and/or their subdivisions?
- Will the reuse require significant changes in the building system? Is it possible to preserve the "patina" of the building?

- Will the reuse respect the value of the cultural heritage and the importance of the site?

Optimising the use of abandoned industrial buildings requires a detailed evaluation of many aspects related to their values (historical, cultural, social, economic, and technological) and existing condition, such as structural design, their ability to accommodate new uses, their potential to meet current standards, the condition of the equipment, the stability of the structure, and the characteristics of the envelope. Successful reuse depends on a combination of factors, including location, size of the site, and the ability of the buildings to adapt to other activities. According to Shipley, Utz, and Parsons (2006), the first requirement for reuse of abandoned industrial sites is a very bold developer. "The existence of dynamic, risk-taking and creative investors who have a passion for beautiful old buildings is probably the single most important element in the industry for the development of cultural heritage". Stratton (2000) notes that there are two basic methods for determining the potential for adaptive reuse: "sixth sense" intuition (most work on adaptive reuse mentions the need for a vision to identify the potential of an abandoned industrial site) and detailed feasibility studies. The study typically addresses the following issues:

Location—Industrial buildings are often located in areas where land values have remained so low that they cannot be sold or demolished. Many of them were located near resources such as water, raw materials, transportation, markets, and labour. Today, these sites are located within the urban area, so their reuse can be linked to smart growth strategies and provide an important benchmark for the community. Typical issues associated with the site include site degradation, or environmental pollution, safety issues, lack or depletion of specific resources, heritage or planning constraints, and an increase in the value of the area sufficient to encourage rapid regeneration independent of the heritage assets.

Condition—The condition of an industrial building, such as its structural integrity, degraded materials, or elements that are difficult to adapt, play an important role in the choice of reuse. A detailed investigation by experts is necessary to assess not only the existing conditions, but also the possibility of additional stresses during and after reuse.

Constraints—Different boundary conditions lead to very specific selection, design and approval procedures. A preservation program not only for the limited parts but also for the architectural values is an important step in the feasibility study. Planning regulations and specific guidelines dictate rules that are essential for the choice of new features.

A methodological approach, as defined by Roviglio (2016), could include the following steps:

Potential Estimation

- Use available tools, such as utility mapping, to identify potential reuse opportunities.

- Potential risks, especially contamination from previous production, should be carefully considered. Decontamination of the site could prove to be an unacceptable cost for many reuses.
- Each disused industrial site certainly has its own characteristics, which need to be carefully assessed through SWOT analysis to understand how they will impact on future reuse. The site, as in any design process, plays a central role in terms of accessibility, services, links and visibility.
- Another important assessment concerns the potential use of existing buildings. It is the load-bearing capacity of structures that makes old industrial buildings attractive for many different functions. Spaces can be easily subdivided and reconfigured according to user needs.

Imagining the Future and Involving Partners

- Design a positive vision for the future in line with local and global trends and interests. Once the vision is shared, partners must be recruited who are able to turn the vision into reality.
- To maximise the chances of success, it is necessary to involve partners at all levels, from citizens to government. Successful projects must be able to convince the first users, such as start-ups, artists, innovators, creative companies, young families, and welcome newcomers, large companies and investors.
- Exchange must be integrated into the design process, making use of the availability of existing space. The activities will facilitate the understanding of future scenarios, generate creativity and show that reuse is already underway.
- Once the reuse process has begun, a more comprehensive design approach is required to create a successful enduring reuse. In the initial phase of the new configuration, an open plan, a clear vision of the goals, and a project that is not yet fully defined can attract new and unexpected players, allow for incremental development, and keep the process flexible and reversible.
- In the project of reuse, the given situation is the starting point, which should encourage and not hinder the process. To make the most of this potential, the project must find the best balance between uses, users and spaces.

Funding and Financial Feasibility

- Feasibility should include initial costs and a realistic forecast of cash flow during operation. The reuse of a site is often a lengthy process.
- Events can be a good way to increase initial funding and generate interest in the project. Cultural manifestations, art installations and temporary markets are all activities that can be organised by the local community to temporarily occupy the site.
- Long-term funding opportunities and public–private partnerships can be a good strategy.
- The incremental process typical of interventions involving the reuse of former industrial sites can encourage forms of funding that have been deferred over

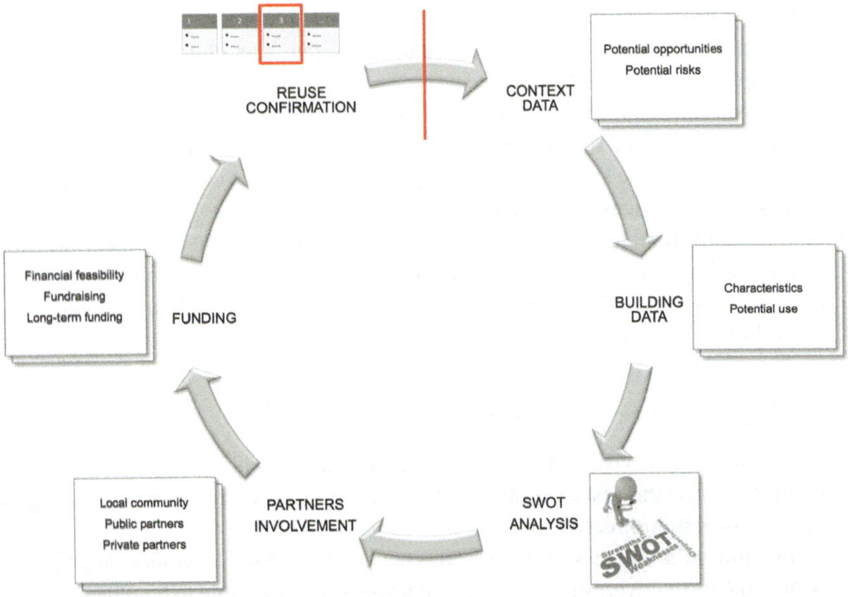

Fig. 4.1 Potential adaptive reuse process. *Source* The author

time. Existing infrastructure can be activated in stages without the need for major investment in the initial phase of commissioning (Fig. 4.1).

4.5 Strategies for Building Renovation

There are many examples of successful renovation of industrial heritage. It is important to emphasise that the concept of industrial heritage encompasses a number of different types, not just buildings exclusively for the productive world. Thus, old demolished railway stations, warehouses, docks and viaducts, to name but a few, become newly adapted buildings, chosen not only for their architectural qualities or historical memory, but often for their strategic location in the city.

In the redesign of industrial buildings, the relationship between the original and the new architecture must be balanced and ensure that the existing constructions can be read (Douet, 2012).

From the numerous case studies examined (only a few are mentioned here), various adaptive reuse transformation tactics can be derived, conceptually schematised in Figs. 4.2 and 4.6. In particular, in the first case, strategies are presented for the many large-volume buildings typical of heavy industry, with large machinery, warehouses or railway stations.

The original volume is considered as a protective envelope, including new volumes and modified to allow natural light and ventilation. Alternatively, it is extended and

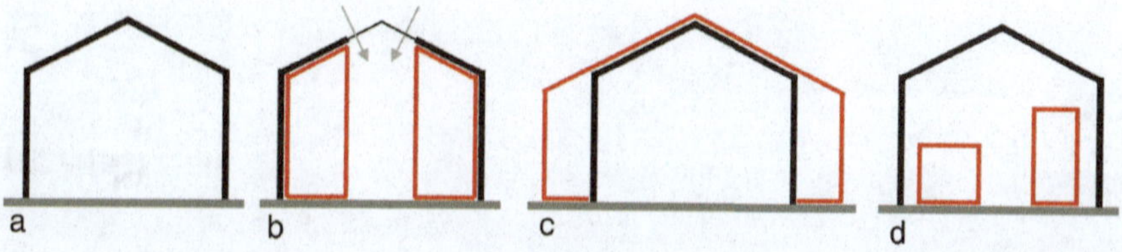

Fig. 4.2 Schematic examples of strategies for reusing industrial heritage. **a** Existing building. **b** Solution with a central atrium providing a lighted buffer zone. **c** A new envelope provides an increase in volume. **d** New volume within the existing envelope. The old building becomes a kind of covered space. *Source* The author

complemented by new parts that interact with the old building to form a unit congruent with the new functions.

This is the case, for example, with the FRAC (Fonds Régionaux d'Art Contemporain), designed by architects Lacaton & Vassal as part of the renovation of the port of Dunkerque (France), which preserves the original warehouse and houses a new, apparently identical, flanking volume with complementary functions. The architects chose not to touch the existing "cathedral", leaving it empty to be used for temporary large-scale events, and combining it with another building, identical in shape and size, into which storage and exhibition areas can be inserted. The new building contrasts the concrete mass of the old shipyard with a lightweight structure clad in corrugated polycarbonate sheets and ETFE 'cushions'. The structure of exposed reinforced concrete beams and columns creates a free and flexible space, as required by the functional briefing.

A significant example of small volumes within an industrial heritage can be well represented by "Casa Mediterraneo" in Benalúa (Spain), designed by Manuel Ocaña del Valle (Fig. 4.3). An old abandoned railway station has been transformed into the headquarters of "Casa Mediterraneo", a non-profit organisation that promotes relations between the countries of the Mediterranean region. The huge old volume, minimally restored, houses many temporary functions and permanent activities are divided into small transparent volumes.

Two examples were chosen to illustrate the reuse of infrastructure (Fig. 4.4). The idea of renovating service facilities with no architectural value often stems from their size and strategic location, which can initiate a process of revitalisation of abandoned and certainly socially degraded places.

Im Viadukt, the EM2N-designed viaduct in Zurich (Fig. 4.5), reuses a derelict part of the city and transforms it into a vibrant centre. Light rail embankments are typically a problem because they cause inevitable fractures in the city, and when abandoned they often become a dangerous place. A famous case was the reuse in Paris (Issy les Moulineaux), designed by architects Dubosc and Landowski, who in 2001 transformed "Les Arches" into an artists' and sports centre.

The Kraanspoor (Fig. 4.6) along the river Ij in Amsterdam (Netherlands), designed by OTH Architects, represents an extreme case of adaptive reuse. The old shipyard

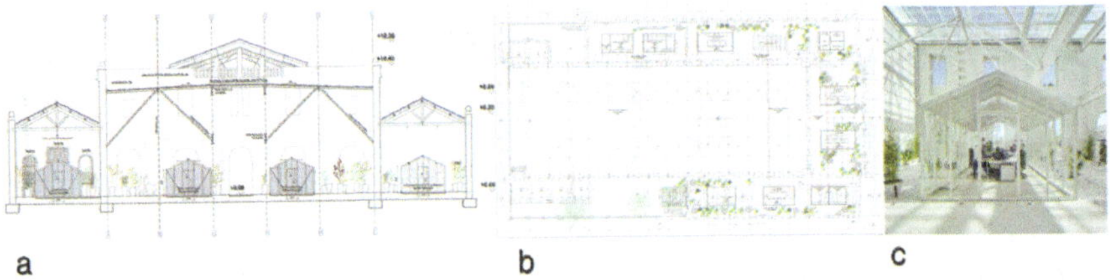

Fig. 4.3 Casa Mediterraneo. **a** Cross-section. **b** Ground floor. **c** One of the permanent boxes. The old Benalúa houses spaces for different events such as exhibitions, concerts, shows or parties, but also administrative offices. After a careful analysis of the building to highlight its value and eliminate surpluses in the face of a reduced budget, some radical changes were made that did not detract from the original charm of the building. The old platform hall remains the main space; it plays the role of a covered square around which small boxes for permanent activities are placed, alternating with flexible and relaxing spaces. The roof has been replaced with translucent panels to ensure maximum lighting. The original windows are removed to increase ventilation and emphasise the effect of an open public space. The small pavilions located along the perimeter are equipped with light and transparent walls, opaque roof and air conditioning to ensure good comfort. *Source* Manuel Ocaña del Valle Architect

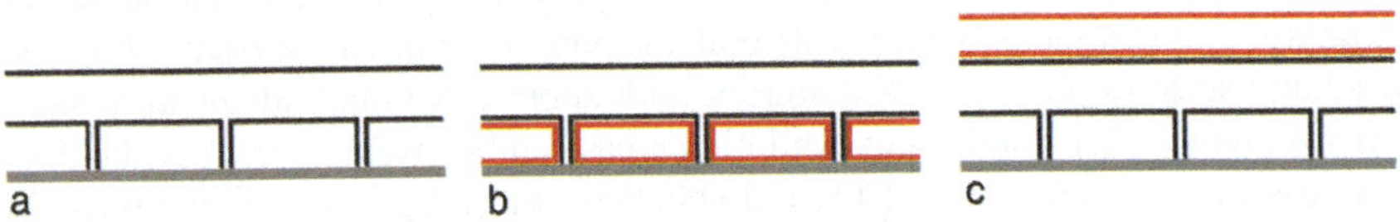

Fig. 4.4 Schematic examples of reuse of decommissioned infrastructure. **a** Existing infrastructure. **b** Newly inserted volume. The example of railway viaducts is typical. **c** Existing structure carries a new volume. *Source* The author

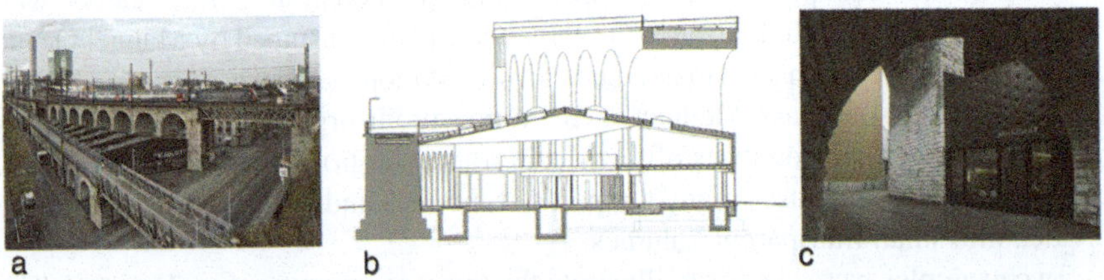

Fig. 4.5 Im Viadukt. **a** General view. **b** Cross-section. **c** Internal view. From an urban obstacle to a connecting structure: A listed railway viaduct near Zurich Central Station was transformed into a 500-m-long shopping mile called "Im Viadukt". A pedestrian and cycle path crosses the lower level of the two railway lines and partly over the new areas. The contrast between old and new is obvious: the thick stone walls remain visible, while new light structures blend into the arches and form triangular volumes between the two railway embankments. Black EPDM roofs with opening skylights characterise the pavilion shells, which are clearly visible from the cycle path. Source Photo **a**, photographer Ralph Hut. Image **b**, EM2N. Photo **c**, photographer Roger Frei

Fig. 4.6 Kraanspoor. **a** The abandoned crane. **b** Waterfront façade. **c** Cross-section. The Kraanspoor (crane runway) is a 270-m-long building on the north bank of the river Ij, twenty minutes from the centre of Amsterdam. It stands on the site of the former NDSM shipyard, which was abandoned in the mid-1970s. The project is quite simple: a large transparent cuboid (13.80 m wide and 10.60 m high) resting on the old concrete structure and supported by a steel frame with 3-m-high pillars. A transparent, double-skin glass characterises the new volume. This climate façade allows natural ventilation of the offices and serves as a buffer against heat in summer and cold in winter. The water of the river Ij is used for heating and cooling. *Source* OTH Architects. Photographer Christiaan de Bruijne

crane was considered solely because of its prominent location and provided the foundation system for the new office building built on top of it.

When renovating former industrial buildings, while respecting and preserving the cultural heritage, the issues of energy saving and comfort required today cannot be ignored.

Energy efficiency measures generally contribute to significant energy savings. Therefore, building envelope improvement strategies should be considered before adaptive reuse of the built heritage.

The most important prerequisites for the design of energy efficient buildings are good thermal properties of the building envelope. Significant reduction in energy demand is possible if a multidisciplinary approach ensures high quality, targeted and tailored energy efficiency solutions.

Energy efficiency strategies can also vary greatly due to the characteristics of buildings, which are often of considerable size. The following figures schematically illustrate some concepts. Particularly for "cathedral" buildings, as they are often defined, an interesting solution may be to downgrade the building envelope to simple protection and create new high-performance volumes inside for permanent activities. The old building will assume the role of an envelope that guarantees the comfort conditions created by a passive approach that varies according to the season—a buffer zone that, if well designed, will provide a milder microclimate than the exterior (Figs. 4.7 and 4.8).

To summarise the concepts expressed, an interesting case is the example of "Halle Pajol" (Fig. 4.9), designed by JAP—Jourda Architects. It is a case of neighbourhood regeneration through the adaptive reuse of a huge warehouse. Two elements make the real value of this type of building: its enormous size and its steel structure, a shell that protects the new volumes underneath. When city officials anticipated demolition,

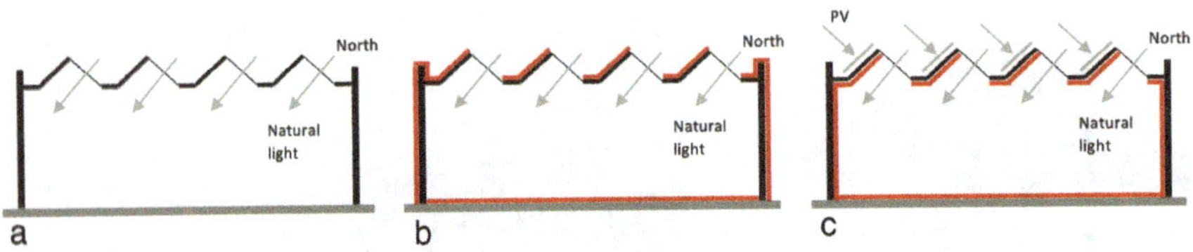

Fig. 4.7 A schematic example of building energy retrofit. **a** Existing building. **b** External envelope upgrade. **c** Internal envelope upgrade. *Source* The author

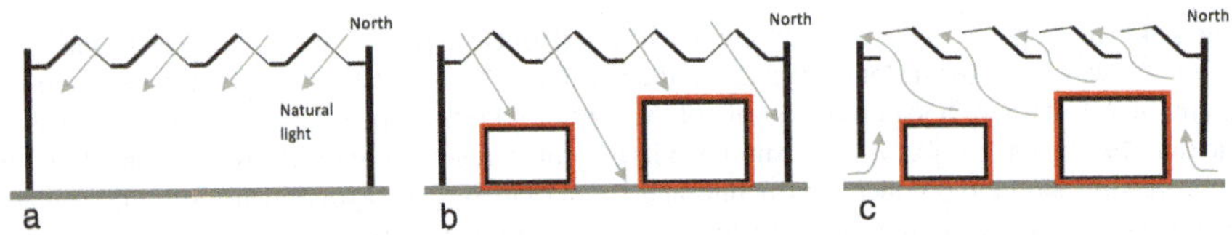

Fig. 4.8 Schematic examples of passive design strategies for buildings. The existing building is considered as an envelope that forms a protected buffer zone. New volumes are designed to ensure a high level of performance. **a** Existing building. **b** Winter set-up. The sun from the south side provides free energy that is absorbed by the floor and walls. **c** Summer set-up. Natural ventilation through the stack effect reduces the indoor temperature. Shading systems are required. *Source* The author

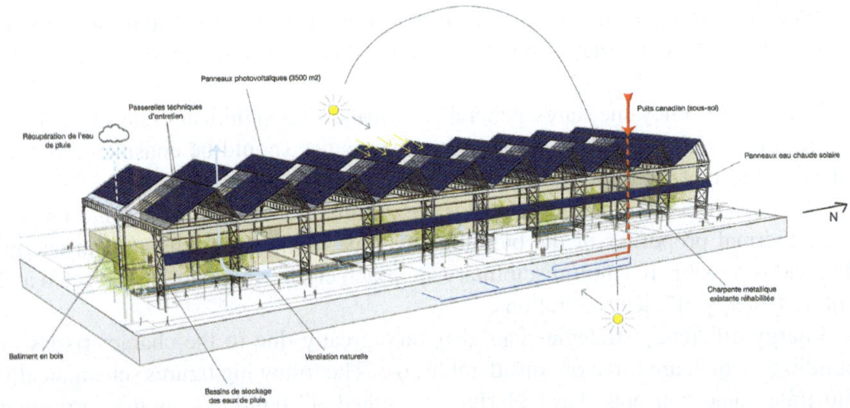

Fig. 4.9 Sustainable schematic design. The renovation of the Halle Pajol old railway warehouses near the Gare de l'Est in Paris is an ambitious and complex project. In fact, architect Françoise Hélène Jourda has succeeded in transforming and completely renewing the neighbourhood by renovating and restoring a steel structure, thus improving the environmental impact and the quality of life of its citizens. In 2007 she won first place in a competition organised by the city council, proposing a solution that preserved the spirit of the site. A large compact volume, inserted between the original steel structures and clad in untreated larch boards, houses a mixture of functions and public covered gardens. *Source* Jourda Architects Paris

residents began advocating for the building's preservation. Key sustainability goals guided all decisions: reduction of grey energy, energy generation from renewable sources, collection and recovery of rainwater for green irrigation, with a consequent reduction of water consumption.

4.6 Conclusion

The reuse of disused industrial sites offers the opportunity to reclaim parts of the city and introduce new functions capable of attracting and virtuously activating an urban renewal process. Revitalisation is an opportunity of great interest today, as the obvious economic and social consequences inevitably represent opportunities for sustainable urban development.

Planning with a careful approach to environmental sustainability can also add significant value to urban regeneration. Optimising the use of disused industrial buildings requires a detailed assessment of many aspects related to their historical and cultural values, but also to social and economic perspectives, taking into account their technical characteristics and their conservation.

In the redesign of industrial buildings, the relationship between original and new architecture must be balanced and ensure the legibility of the existing structures.

The transformation project must carefully consider which passive and active strategies best ensure the achievement of the energy and performance targets required today.

References

Banham, R. (1986). *A concrete Atlantis: U.S. industrial buildings and European modem architecture 1900–1925.*MIT Press.

Douet, J. (Ed.). (2012). *Industrial heritage Re-tooled: The TICCIH guide to industrial heritage conservation. TICCIH*. Carnegie Publishing.

Falconer, K. (2005). Industrial archaeology goes universal. *Industrial Archaeology Review, XXVII*(1).

Kincaid, D. (2002). *Adapting buildings for changing uses: Guidelines for change of use refurbishment*. Taylor and Francis.

Landorf, C. (2009). A framework for sustainable heritage management: A study of UK industrial heritage sites. *International Journal of Heritage Studies,* 15(6).

Oevermann, H., Degenkolb, J., Dießler, A., Karge, S., & Peltz, U. (2016). Participation in the reuse of industrial heritage sites: the case of Oberschöneweide, Berlin. *International Journal of Heritage Studies, 22*(1).

Robiglio, M. (2016). The adaptive reuse toolkit. How cities can turn their industrial legacy into infrastructure for innovation and growth. In Urban and regional policy paper, n. 28, The German Marshall Fund of the Unites States.

Shipley, R., Utz, S., & Parsons, M. (2006). Does adaptive reuse pay? A study of the business of building renovation in Ontario, Canada. *International Journal of Heritage Studies, 12*(6).

Stratton, M. (2000). *Industrial buildings: Conservation and regeneration*. E&FN.

Vardopoulos, I. (2019). Critical sustainable development factors in the adaptive reuse of urban industrial buildings. A fuzzy DEMATEL approach. *Sustainable Cities and Society, 50.*

References

Chapter 5
Building Technical Upgrade

Abstract Before starting a renovation project, a careful analysis of the building's characteristics is necessary. Knowledge of the specific building technologies and their critical details will help to better identify the causes of deterioration and possible remedies. Typically, a renovation project involves major changes to the existing structure, including structural interventions that require careful engineering evaluation. Alignment with current building codes and the need to improve performance are other important aspects to consider in a renovation project.

Keywords Building technologies · Repair techniques · Performance upgrade · Refurbishment

5.1 Introduction

The essence of a restoration process is the preservation of the building, regardless of its possible reuse. It is not for nothing that we speak of "conservation", the main aim of which is to repair decay or remove work that has led to a change in the original historic value. Similarly, the process of building renovation and adaptive reuse must well assess the condition of the building and define any work necessary to remove deterioration.

As mentioned in Chap. 2 (see Box 2.2.2), it is necessary to establish a correlation between deterioration and its causes before planning maintenance work. Buildings develop a variety of problems over the years. Choosing an appropriate repair method requires systematic observation of symptoms and their possible causes, and requires a high level of expertise. Sometimes the process is not so simple and linear because it is necessary to understand what mechanisms have caused the failure, what external factors have created a "disease state" in the building.

A diagnostic process follows the typical steps of a doctor: it involves a series of examinations, analyses, tests and findings that lead to a diagnosis of the causes of the "disease" and a "medical report" with a hypothesis of cure.

A good designer, like a good doctor, is not necessarily an expert in each discipline, method and testing tool, but must be able to manage the whole process by using experts in the different phases.

M. Grecchi, *Building Renovation*, PoliMI SpringerBriefs,
https://doi.org/10.1007/978-3-030-89836-6_5

Each phase requires a specific project to determine what the key actions are and how they will be implemented. The systematisation of many specific topics is fundamental: chemistry, physics, knowledge of materials and their behaviour, physical and technical processes of heat transfer, condensation, data collection from context, construction techniques and construction behaviour, to name the most important.

Knowing what construction techniques were used in the period and location of the building allows important evaluations to be made before sophisticated and costly investigations are planned. Each historic period has produced typical construction methods and details that result in deterioration or weaknesses that must be addressed, regardless of how the building is renovated. The more complex the technique, the more difficult it becomes to understand the causes that led to deterioration.

Systematising all interventions to address deterioration, performance improvements and regulatory adjustments with the functional program allows coherent decisions to be made in the renovation project.

5.2 Criticality of Construction Technologies

Another important factor in favour of renovating old, redundant, or obsolete buildings is that many of them are well built and structurally sound. However, a detailed study is required to confirm their quality and condition in order to determine the expected cost of the necessary adaptations and their impact on the feasibility of the renovation program.

Before starting the renovation of the building, it is necessary to carefully study the characteristics of the main parts of the structure, e.g. façades, roofs and panels, as these are typically the elements to be preserved and adapted.

Many existing buildings have problems related to ageing of materials, degradation due to various factors, and also construction methods that have inherent weaknesses. Each era has produced typical building technologies, some of which are strictly local.

Boxes 5.2.1 and 5.2.2 give examples of some typical European building technologies, highlighting their evolution and the most critical details that require careful study.

5.2.1 Box: Exterior Walls

The evolution of building techniques analysed from the nineteenth century onwards, partly because most of the buildings considered in the renovation are not architectural monuments, can be judged against significant historical thresholds. This change has often led to experimentation with new solutions, sometimes making mistakes due to a lack of expertise or poor technical evaluation of the earlier building systems.

5.2.1.1 Technology Up To 1940s

Solid Walls

Load-bearing brick walls can be considered a "single-layer" system, even though there are typically interior and exterior face layers. At the same time, they guarantee all the performances required for a vertical finish. As far as waterproofing is concerned, critical problems can arise from the direct transfer of moisture from the exterior to the interior, exacerbated by plaster and deterioration of mortar joints. Masonry walls are load-bearing structures made of clay bricks or stone blocks, typically protected with cladding layers (plaster, natural or artificial stone slabs, etc.).

Arrangement and overlap of bricks—The way the bricks are arranged in the wall is directly related to the method of construction used and can be a weak point in the system. Solid walls often vary in thickness, decreasing from bottom to top and thinning under windows; they often contain chimneys and ducts (Fig. 5.1).

The typologies of masonry depend mainly on the characteristics of the blocks (brick or stone) and the joints. The materials used for the joints, which represent 20–30% of the total surface area of the wall, are very important; they allow the wall to withstand normal stresses and thermal movements.

Stone walls are generally made from local materials ranging from limestone and flint to granite and sandstone. However, the performance of stone can vary greatly due to its resistance to weathering and water penetration, and the ability to be chiselled into regular blocks.

Cladding—The protective layers were often made of plaster and artificial stone. Cost-effective buildings with adequate decoration were the main objective: some technological solutions show a "transfer" of the same details used for natural stone to these new prefabricated elements. The artificial stone was made with a mixture of

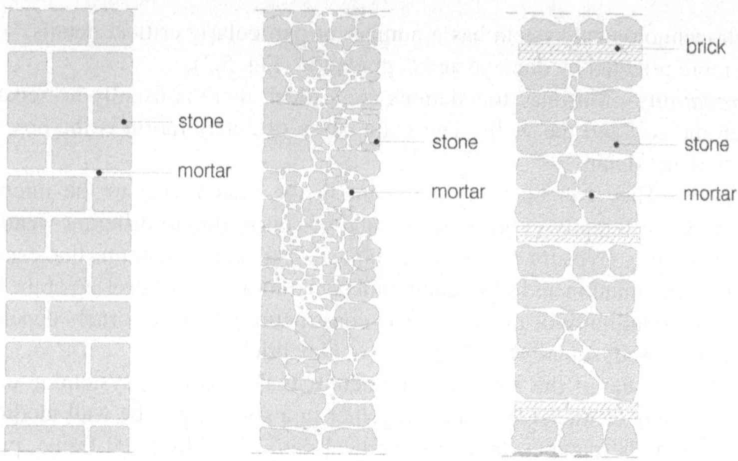

Fig. 5.1 Typical stone walls

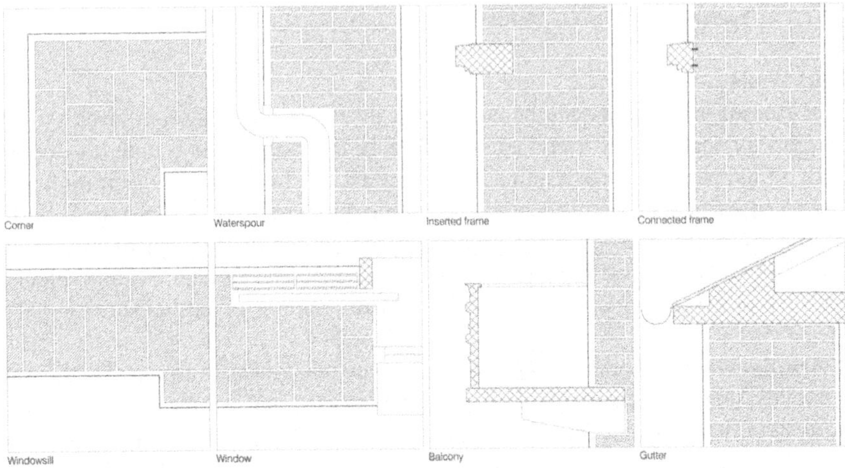

Fig. 5.2 Crucial details

cement and stone grains of different grain sizes and colours to give the joint a natural appearance.

Crucial Details

Frames and claddings are very important for the performance of façades:

- The surface of the render, delimited by horizontal and vertical frames, reduces the effects of stresses due to sudden changes in temperature and prevents cracks in the cladding.
- Frames made of natural or artificial stone and windowsills reinforce and protect the weakest parts of the façade.

Each technological system has a number of particularly critical details, so they can be more affected by damage and degradation (Fig. 5.2).

Foundations—Masonry foundations are critical; there is usually no separation between the soil and the walls. The main cause of deterioration is the process of capillary rising damp.

Corners—They can be a critical point in the system due to the interaction between two differently exposed walls (and of course due to different weathering and phenomena). Gypsum coatings are subject to stresses and strains that can cause cracking, delamination and subsequent rain penetration. Proper overlap of the bricks must ensure continuity of the walls. The construction of stone corners depends on the materials used: square stone elements or clay bricks.

Windows—One of the most delicate details is the masonry system around the windows. Many elements are joined together in a small area: the wall modified in different sections (parapet, lintel and vertical parts), cladding and frame, parapet, windows, sunshade system. There is a succession of small horizontal and vertical surfaces that are highly exposed to weathering.

Frames—Three typical ways to build frames are:

- Masonry and cement conglomerate form, often with the addition of marble or stone grains applied in successive layers with special tools.
- Prefabricated elements of considerable thickness, which are inserted into the walls during construction. The solution is similar to the one used for shaped elements of natural stone.
- Prefabricated thin elements that are fixed to the wall with iron clamps.

Degradation Phenomena

Mechanisms acting on masonry, such as physical changes (stress due to loads, thermal movements, efflorescence due to frost, alveolar erosion) or chemical changes (carbonation, sulphation and oxidation phenomena) can be motivated by intrinsic or extrinsic causes. Among the latter, all phenomena related to moisture are the most important.

Excessive Moisture

Excessive moisture, caused by rising damp and water ingress, is one of the most damaging causes of building decay. When combined with temperature variations, it has a greater effect on the overall performance of materials. Most defects are caused and aggravated by the presence of moisture.

Rising damp—Moisture increases within a wall by capillary action due to natural osmosis, which causes water to move from a lower solution to a higher salt concentration. This depends on the size and distribution of pores in the masonry, the presence and concentration of soluble salts, and the electrical potential of the wall relative to the surrounding soil. Jurin's law describes the rise and fall of a liquid in a thin capillary tube. The presence of excessive moisture causes damage to walls and surfaces and accelerates the crystallisation of salt and the effect of frost (Fig. 5.3).

Penetration of moisture—Rain, increased by strong winds, may enter the wall through the mortar joints. The movement of moisture absorbed in a porous material depends on the severity of the exposure conditions, the duration of the phenomenon, and the internal structure of the material pores. Damaged gutters, downpipes and installations are another source of moisture.

Surface and interstitial condensation—Relative humidity is inversely proportional to air temperature, i.e. when the temperature falls, relative humidity rises to the condensation point (dew point). When the air is saturated (100% RH), water condenses on cold surfaces (surface condensation). When the temperature of the dew point is reached within the wall thickness, condensate forms there (gap condensation).

Movements

Differential motion, over homogeneous sections or along the boundaries of two different elements or materials, produces stresses and causes degradation phenomena.

Thermal fluctuation—Materials (except water) expand when heated and contract when cooled. The measure of material movement is called the coefficient of thermal expansion.

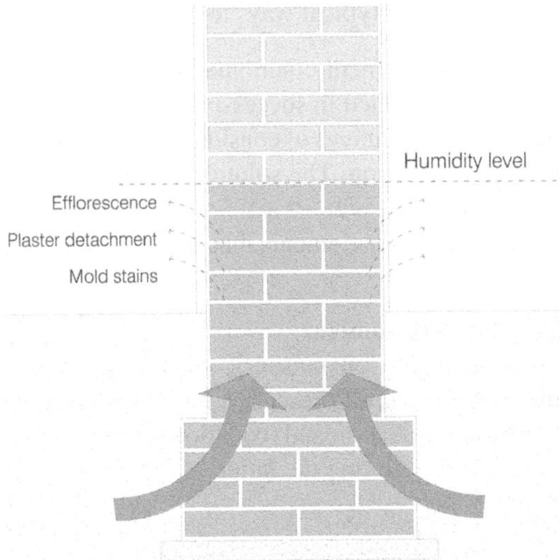

Fig. 5.3 Jurin's law is expressed as: $h = 2\gamma\cos\theta/r\rho g$, where: h is the liquid height; γ is the liquid surface tension; θ is the contact angle of the liquid on the tube wall; ρ is the liquid density; r is the tube radius; g is the gravitational acceleration

When materials are subjected to cycles of high and low temperatures, they undergo cycles of expansion and contraction. For a homogeneous material (e.g. stone), this cyclic movement between the surface and the body can result in damage and detachment of the surface layer. When materials with different coefficients are coupled, the difference in expansion and contraction leads to damage of the weaker material.

Similarly, stresses generated within a material that cannot expand and contract, e.g. due to incorrect fastening systems, lead to cracks.

Hygroscopic variation—Water absorption typically leads to an increase in the volume of the material. The major difference between thermal and hygroscopic expansion is their relative degree of predictability and reliability. Hygroscopic expansion is usually expressed as a percentage of the original size. Porous masonry, such as bricks and stones, can be subject to thermal and moisture movement, which means they are particularly susceptible to deterioration and decay.

Soil properties—Soil properties and various factors such as landslides and changes in water levels can cause movement and instability. Foundations of existing buildings are often inadequate in relation to applicable standards and may have been subjected to increased loads due to changes in their purpose or design changes.

Settlements—Movements within a structure due to the distribution of loads and stresses between the various elements of the structure usually occur during the first few years of the structure's life. Settlements caused by changes and redistribution of loads lead to cracks and instability.

5.2.1.2 Technology After 1940s

Follow Brick Walls

The typical method of hollow brick walls was introduced in the late nineteenth century in north-western Europe and has been widely used since the 1920s.

The spread of reinforced concrete construction determined a significant development in façade design. Starting from rationalism, the distinction between load-bearing construction and curtain walling led to a loosening of the envelope system and the abolition of all ornaments, which were considered only as decorative details and not as technical protection.

A critical analysis of the first solutions from this period shows the simplicity of the details, which was partly due to strong architectural thinking that was not supported by adequate technical control.

The double-walled system using hollow bricks represents the first step towards performance design.

The widespread adoption of cavity insulation began in the 1970s (first oil crisis and mandatory standards for building insulation) (Fig. 5.4).

Crucial Details

Interaction between reinforced concrete structure and brick walls—One of the most important details is the connection between reinforced concrete structures, columns and beams and curtain walls. Two main elements need to be checked:

- homogeneity and continuity of the wall to ensure that the cladding is independent of thermal or settlement movements of the different materials (concrete and brick),
- in the presence of insulation in the cavity, thermal bridging phenomena corresponding to the concrete structure not protected by a suitable insulation layer (Fig. 5.5).

Cladding systems—Exterior claddings vary in materials, construction methods, and finishes. In general, many stresses and resulting delaminations are created by connections between the wall and cladding.

Rendered surfaces are critical when there is:

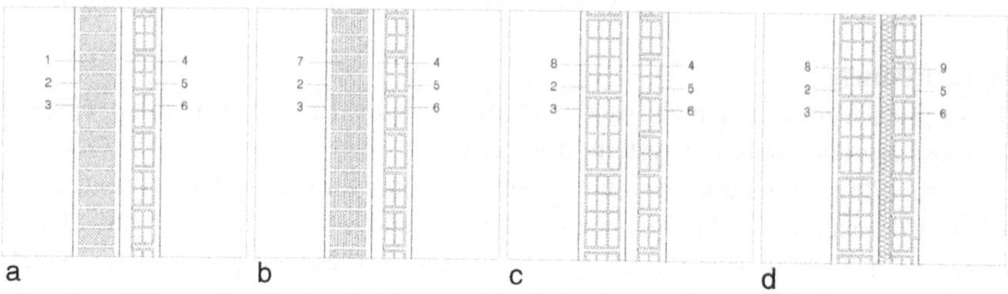

Fig. 5.4 Multi-layer walls, different solutions: **a** External wall with solid bricks. **b** External wall with vertical hollow bricks. (3) External wall with horizontal hollow bricks. **d** Since 1970s cavity with thermal insulation. (1. Solid bricks. 2. Plaster. 3. Cladding. 4. Cavity. 5. Hollow bricks. 6. Plaster. 7. Semi-solid bricks. 8. Hollow blocks. 9. Thermal insulation)

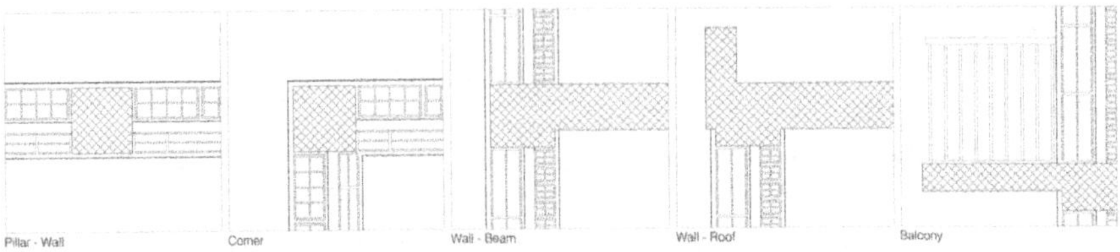

Fig. 5.5 Multi-layer walls, Concrete structure and brick walls interaction

- excessive elongation compared to the elasticity of the mortar,
- a discontinuous wall that can transfer stresses to the coating, resulting in cracking and delamination,
- insufficient adhesion between substrate and plaster, which can lead to bulging and delamination.

Stone tile systems depend on the size, thickness and properties of the stone. Critical details are:

- if the weight of the tile is significant, the installation method must provide stability and differential movement between the substrate and the cladding. The best systems use mechanical joints,
- joints when tiles are fixed with mortar.

Glass tiles and small ceramic tile coverings, typically used in the 1960s, have been phased out due to inadequacies in installation methods and elements that are not suitable for atmospheric loads. Detachment can be caused by:

- mechanical stress from the combined action of wind and rain, resulting in damage to joints and water ingress, which can reduce the adhesive's resistance,
- aggressive weather conditions, such as acid rain and air pollution,
- improper installation of ceramic elements, combined with differential thermal expansion between elements, joints and mortar surfaces,
- defects in the production and installation systems. This is the case of glass tiles that have been emptied onto sheets of paper (approx. 60 × 60 cm), placed on a mortar surface and then washed to remove the paper.

Precast Panels

Heavy precast concrete panels, used in many sprawling European suburbs, represent the most degraded building system of the day.

In the 1960s, companies began to develop patents for precast panels coated with reinforced concrete, thermal insulation, exterior and interior cladding, and characterised by a modular size (typically 3.50 × 3.00 m). Façade panels, floor panels, interior partitions and staircases were made with the same precast system, resulting in rigid and problematic functional divisions.

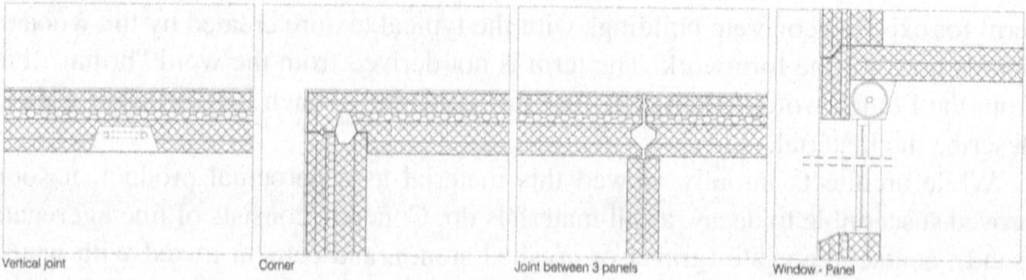

Fig. 5.6 Typical precast panel systems used in the 1960s

The post-war reconstruction and the new industrial expansion, which led to typical migratory phenomena, required the ability to construct many square metres of residential buildings in a very short time. The prefabricated panel system was particularly suitable for social housing.

All panels are load-bearing and are hard to remove without affecting the stability of the building. Joints and connections between panels are some of the most vulnerable parts of the system.

All the patents had common shortcomings, mainly related to the size of the panels, the manufacturing process and the reduced thickness, which cannot guarantee the thermal and acoustic performance. However, the undisputed advantages were the acceleration of construction and the immediate response to requests for low-cost housing in large quantities. These methods did not guarantee the expected results and were therefore soon abandoned (Fig. 5.6).

Crucial Details

Today, these buildings are characterised by any kind of obsolescence in terms of envelope performance and the functional layout of the apartments.

Renovation must consider solutions that can simultaneously transform the suburbs by solving social problems, the low level of building performance and updating the functional arrangement that no longer meets the needs of users (see some examples in chapter six).

Joints between precast panels—All the systems used in the 1960s and 1970s had serious workmanship defects, poor insulation, and movement caused by the weight and size of the panels. Mortar joints on the site began to be a problem immediately after construction.

The materials used were not able to withstand stresses, thermal movements and weathering. These phenomena led to the detachment of the protective surfaces and consequently to the penetration of water, causing corrosion of the concrete reinforcement and moisture.

Raw Concrete Walls

In the twentieth century, a wide variety of architecture was built with exposed concrete, both as a practical and economical choice and for its innovative aesthetic qualities. Cast-in-place concrete was easily adapted to modern styles. Brutalism is a term for exposed concrete buildings with the typical texture created by the wooden boards used for the formwork. The term is not derived from the word "brutal", but from the French word "béton brut" or "raw concrete", which Le Corbusier uses to describe the material.

While architects initially viewed this material as a perpetual product, it soon proved susceptible to decay, as all materials do. Concrete consists of fine aggregate (sand), coarse aggregate (gravel or crushed stone), and cement mixed with water. The predominant material by mass is the aggregate. Portland cement is the most commonly used binder in modern concrete.

The quality of concrete depends on the ratio of water to binder. Excessive use of water in the hydration process results in permeable concrete, which is more susceptible to weathering, and deterioration of the layer protecting the steel reinforcement, which causes corrosion.

Crucial Details

Concrete deterioration occurs mainly due to corrosion of embedded steel, material deterioration, use of improper techniques and structural problems.

Outer layer—When the normal alkaline environment of the concrete (above a pH of 10) is compromised and carbonation reaches the metal reinforcement, the concrete no longer protects the steel from corrosion. If the coating thickness is too thin to counteract the expansion of the steel caused by oxidation, the concrete coating will disintegrate.

5.2.2 *Box: Roofs and Floor Slabs*

The main system until the 1940s, with the exception of large industrial or service buildings, was the pitched roof with wooden structures. There was considerable variation in this, determined by the size, complexity and characteristics of the protective layers.

Historically, different climatic conditions and traditional materials have influenced the geometry and type of roof. Tropical regions have produced roofs that are generally very sloped, light, and ventilated to drain large amounts of water quickly and reduce high temperatures thanks to air movement. The most northern regions have produced heavier roofs to withstand the weight of snow and reduce heat loss. Flat, lightweight roofs to minimise solar radiation absorption, with the addition of rainwater collection systems, characterise dry climates. Only temperate regions are not affected by climate change, but have produced solutions based on locally available materials.

Pitched Roofs

There are different roofing systems depending on shape and geometry, building materials, stormwater drainage, building size, and architectural styles.

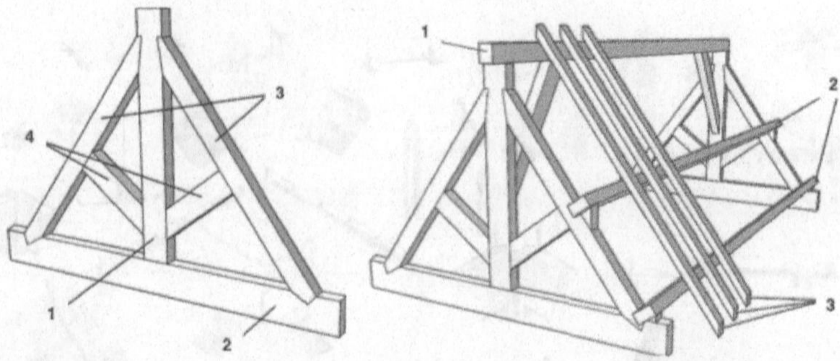

Fig. 5.7 Left, 1. King post—2. Tie beam—3. Principal rafters—4. Struts. Right, 1. Ridge board, 2. Purlins, 3. Common rafters

Traditionally, wood was the most common building material used for roofing (typically in Europe), and the shape and proportions were based on the span of the beams between the columns, usually load-bearing walls.

The pitch of the roof was determined both by the properties of the roofing material and by local traditions, often dictated by weather conditions. Various cladding systems, which must ensure watertightness and withstand wind pressure, require a specific pitch to ensure proper overlap of the elements. Roof pitches can vary from 60 to 30 degrees or less, depending on the different tile types.

When a roof system needs to be refurbished, the main structure is usually carefully identified for possible restoration (Fig. 5.7).

Timber structure—The construction scheme varies according to the distance between the bearing walls, the complexity of the roof geometry, local traditions and wood characteristics (typical in Europe: oak, North Pine redwood, whitewood). It is possible to find very simple solutions consisting of elements resting directly on the walls and not on a ridge beam, or of trusses whose complexity is related to the distance between the bearing walls.

In important buildings, there are structures with complex trusses capable of supporting and redistributing the loads on the bearing walls. A truss is a rigid framework of beams, joists and braces capable of spanning great distances without intermediate supports. Trusses are usually spaced at regular intervals and connected by stringers. Roof structures and trusses were often designed off-site, disassembled, and assembled on-site.

A roof truss has two main rafters, a tie beam at the bottom, and a central vertical king post. The king post is usually in tension and must be properly connected to the tie beam and main rafters. It should never be rigidly connected to the tie beam as it should prevent it from bending and twisting. A metal bar can replace the king post.

The use of iron—Iron was an integral part of the wooden roof truss as a link between the elements. Another development was the use of iron instead of wooden tie rods. From the early nineteenth century, cast iron beams and trusses became common in large industrial and public buildings. From the end of the nineteenth century, steel gradually replaced cast iron (Figs. 5.8, 5.9 and 5.10).

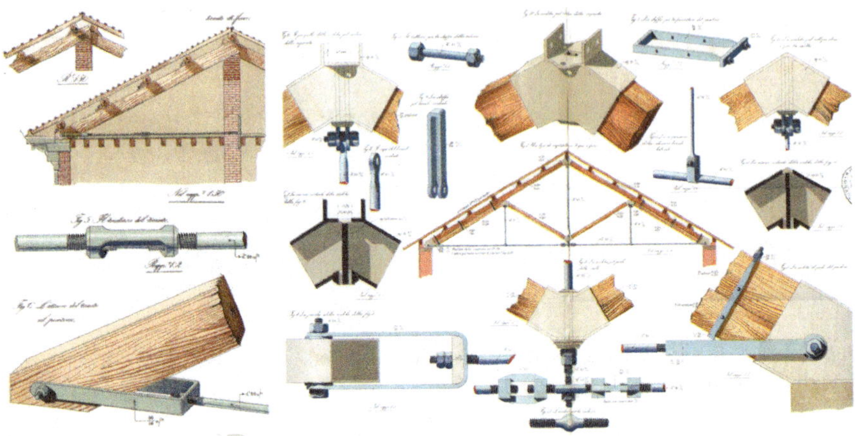

Fig. 5.8 Wood-iron truss systems. *Source* Formenti, C., La Pratica del Fabbricare (1909)

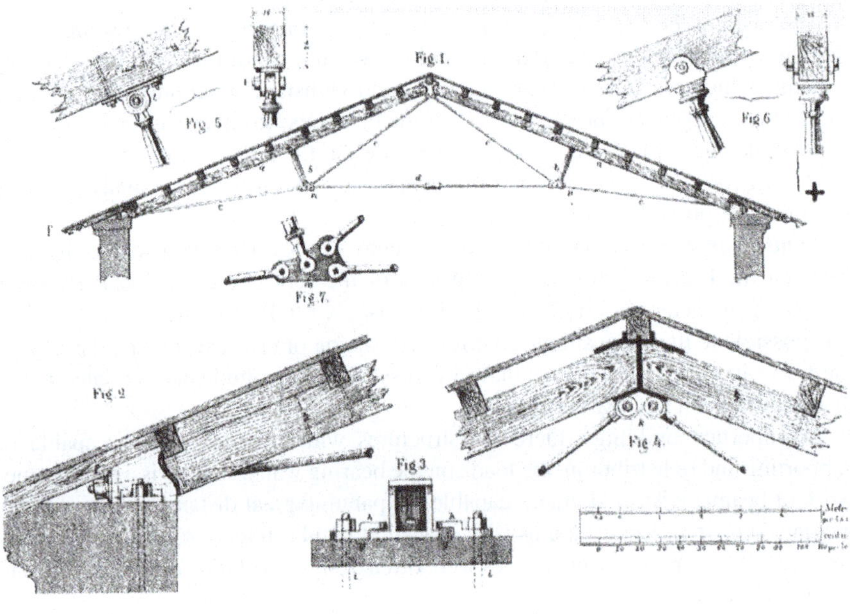

Fig. 5.9 Iron truss systems, Polonceau Patent (1837)

Cladding systems are linked to the local tradition, climate, and available materials.

Clay tiles—The Romans were the first to use clay tiles as a roofing material. The clay tiles, both flat and pantiles (with a double "S" curve profile), typically used in the southern regions of Europe, were attached to the wooden beams, neatly overlapped, and sealed with mortar at the top of the roof to prevent rain penetration and protect

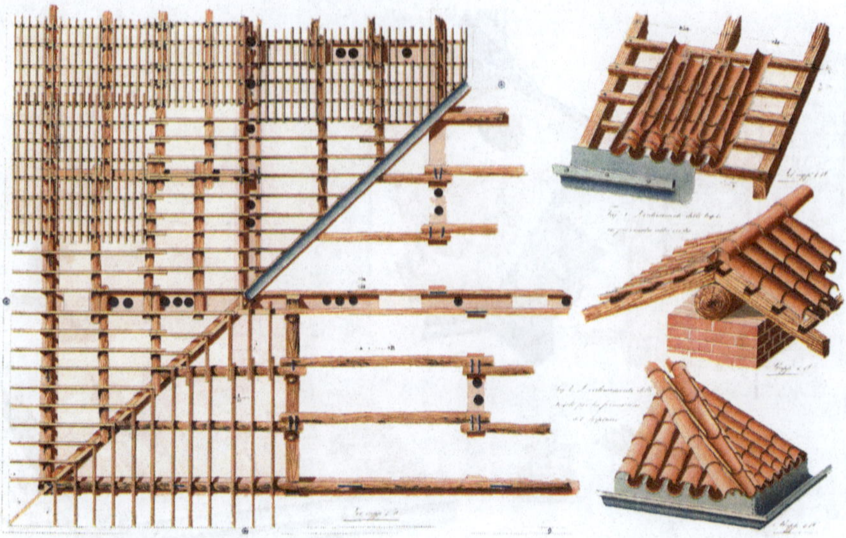

Fig. 5.10 Roof details, clay tiles. *Source* Formenti, C., La Pratica del Fabbricare (1909)

against wind. The roof tiles were nailed every third layer. A ridge element protects against rainwater and allows natural ventilation under the roof.

Slates—The slates, whether small or large, often irregular, were generally fastened to the wooden beams with nails. A ridge element prevented the penetration of rain.

Stone slabs—These were used as roofing material when the right stone was available on site. Where stone slabs were used, more solid and stronger wooden structures were required to support the heavy roof. Each row of stone slabs was laid so that there was sufficient overlap with the layer below, decreasing in size towards the ridge. Because of their size, weight, and thickness, the elements did not need to be attached.

Wooden shingles—The wood shingles used in some northern regions are thin sheds of cedar, oak, or chestnut. This coating requires a well-ventilated system to keep the wood dry.

Lead sheets—In the past, lead was used for gutters, rainwater pipes and fittings, and sometimes for cladding important public buildings, as it was an expensive and therefore valuable material. Some cupolas, dormers and small domes are still covered with lead sheets. These plates were laid on wooden boards and joined by hot welding.

Copper sheets—Copper sheets have been used as a roofing material in Europe since the Middle Ages, especially on some of the great churches and cathedrals. In the twentieth century, copper was used to cover flat roofs and as an alternative to lead for flashing. Traditionally, copper sheets were handmade and sewn together on site. By the end of the nineteenth century, they could be produced in thin sheets of uniform thickness.

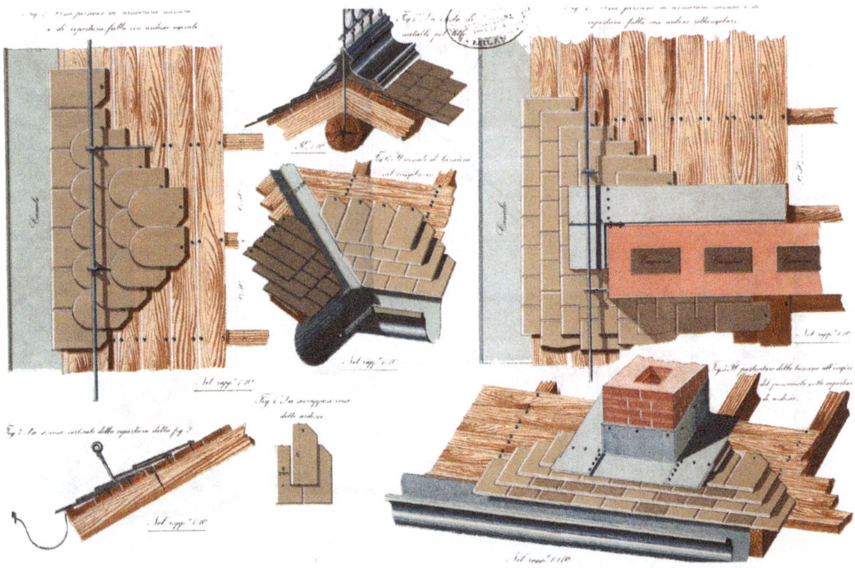

Fig. 5.11 Roof details, stone tiles. *Source* Formenti, C., La Pratica del Fabbricare (1909)

Corrugated metal—Corrugated metal sheets were used to cover large flat roofs or slightly sloped and curved surfaces. The sheets were attached to the structure every four corrugations with galvanised nails (Fig. 5.11).

Crucial Details

Finally, all the accessories of the roof that contribute to its proper functioning are really important: gutters, rainwater, drains, flashing, ridge, impluvium and chimneys are elements of discontinuity that must be carefully monitored, especially to identify the causes of deterioration or failure.

Gutters and downspouts—In the nineteenth century, cast iron replaced lead as the most common material for collecting rainwater. Gutters and downspouts were attached to masonry with cast iron brackets.

Gutters were made in a variety of ways depending on the architectural decoration. Stone frames inserted into the upper part of the wall could conceal the gutters. The wooden beams connected to the joists could cover the roof overhang with the gutter.

Chimneys—Originally built with brick walls and covered with a clay roof, chimneys were placed in various locations for practical and aesthetic reasons. Then, in the twentieth century, prefabricated chimneys made of clay or copper were preferred.

Dormer windows—So called because they originally lighted the sleeping area in the attic, dormer windows were made of wood, with roofs of one or two pitches.

Floor Slabs

Different systems were used depending on local tradition, size and architectural importance.

Their specific characteristics depended on the building material used. Wooden structures were typical of the nineteenth century, when load-bearing walls were the most common vertical support system.

Timber and mixed iron-timber structures—There were very simple solutions, with wooden beams from wall to wall and wooden boards above them as floors. More advanced systems consisted of main beams, cross beams and wooden boards, with floors of different materials.

Mixed wood and iron construction with poured concrete is a further development of the floor slab systems. No system was integrated into the floors (Figs. 5.12 and 5.13).

Arched brick floors—Two main systems were used: arched structures and flat structures. Both systems were built with hollow clay elements of different sizes and shapes, similar to today's hollow bricks. They were built on wooden formwork platforms on which the elements were positioned and then fixed with a concrete pour. Often tie rods held the steel beams together (Fig. 5.14).

Reinforced concrete beams and hollow clay block systems—These systems have been used in some regions since the early twentieth century. The elements were

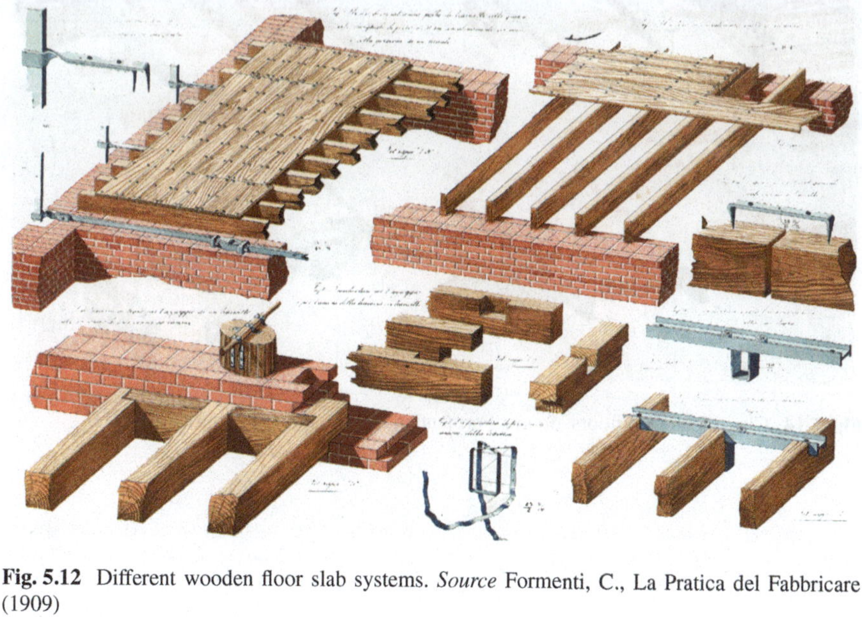

Fig. 5.12 Different wooden floor slab systems. *Source* Formenti, C., La Pratica del Fabbricare (1909)

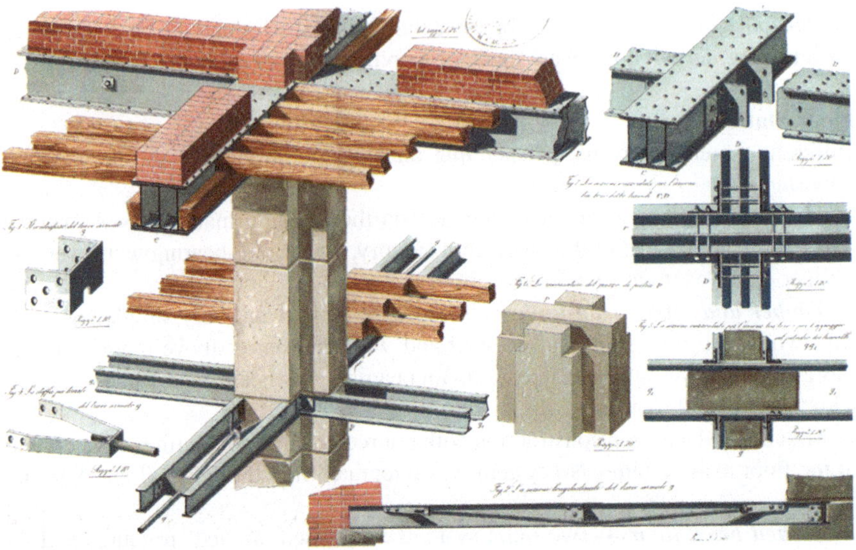

Fig. 5.13 Mixed wood-iron floor slab systems. *Source* Formenti, C., La Pratica del Fabbricare (1909)

Fig. 5.14 Clay-tile arched floors. *Source* Formenti, C., La Pratica del Fabbricare (1909)

positioned to form voids that allowed the reinforced concrete beams to be poured and form a monolithic structure after the top layer of concrete was applied.

Precast concrete beams and hollow block systems—Precast beams and slabs with masonry elements have been manufactured since the 1920s. They represent an evolution of the previous system.

Precast concrete beams or steel girders were placed parallel and at regular intervals, then hollow bricks were placed between the beams and consolidated by pouring concrete. Or prefabricated clay beams were positioned parallel and side by side and connected at the top by a concrete pour. No formwork was required for either system.

Precast reinforced concrete systems have been used for factories since the early twentieth century. The first solutions came from earlier systems (Fig. 5.15).

Flat Roofs

Flat roofs were historically used frequently in arid regions, sometimes as living quarters on cool evenings.

Since the late nineteenth century, when new waterproofing materials were invented, flat roofs have been most commonly used in Europe and the United States for warehouses, office and commercial buildings, and many urban housing developments. The first waterproofing membranes were made of tar or bitumen and had to be protected to prevent damage.

Flat roofs have a number of sensitive technical details and some construction errors are typical. If these details are not well executed, water will certainly penetrate. Flat roofs have the same structural system as the lower floors, but there are different technical solutions. Since the 1970s, an insulation layer has been introduced and its position between the other layers determines three different operating models (Fig. 5.16).

Warm flat roof—The term "warm" refers to the location of the insulation layer in relation to the slab. In a warm flat roof, the insulation layer is above, meaning the structure is on the warm side of the insulation. A vapour-tight layer (located on the warm side of the insulation) is usually required to prevent gap condensation. Depending on the type of roof membrane, a ballast system may be required.

Inverted warm deck flat roofs—Insulation is installed on top of the waterproofing layer. This protects the waterproofing from thermal stress, UV degradation and

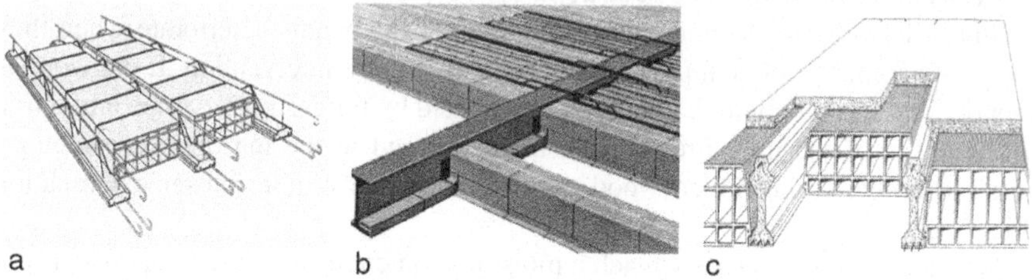

Fig. 5.15 European systems, starting from 1950. **a** In situ solution. **b** RDB patent, Italy. **c** Prefabricated solution

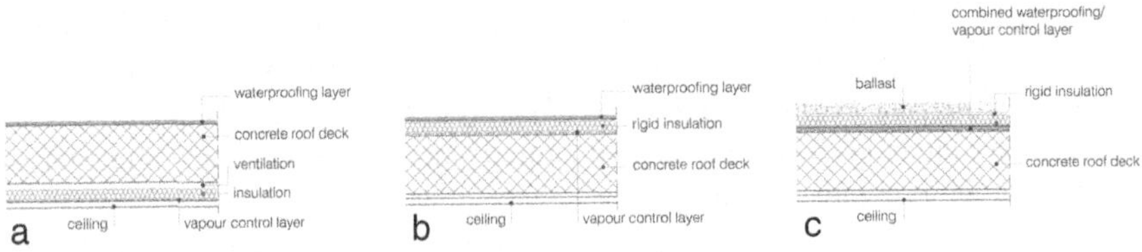

Fig. 5.16 **a** Warm flat roof. **b** Inverted flat roof. **c** Cold flat roof

mechanical damage. No vapour barrier layer is required as the roof membrane also provides this service. A ballast is usually required.

Cold flat roofs—Insulation is typically placed between and/or under the floor slab, which means the structure is cold. A vapour barrier layer is required on the warm side of the insulation to prevent the formation of interstitial condensation. Ventilation can be provided between the insulation and the slab to facilitate air flow.

Roof Decay

Poor maintenance practices are the main cause of roof deterioration. Depending on the type of roof, seasonal and extraordinary inspections are recommended. In the event of deterioration, an engineering inspection is required.

The first step includes a structural inspection (materials, stability and condition) and analysis of all materials and elements, such as cladding, gutters, pipes, chimneys, dormers and skylights.

It is necessary to determine what the possible causes of this specific deterioration are. Particular attention should be paid to areas where water can penetrate; typically, these are joints between different elements, flashings, ridges, tile movement and parapets. Timber structures require detailed inspection to identify fungal or insect infestation, corrosion of metal joints and their structural capabilities.

Bending, twisting, or warping of beams and joists can be caused by a variety of factors, including wood rot, defects in the original construction, increased loading, or serious problems in the load-bearing walls.

Historic roof beams are often older than the roof covering, which may even be replaced several times over the life of the building. Careful examination of roof forms, joint details, and beam features can help date the roof structure.

Decay and failure of wooden structures—Typically, old wooden structures have localised degradation. In many cases the ends of the beams deteriorate when they come into contact with damp external walls and require intervention. If the wood is kept dry and well ventilated, it will not be affected by fungal attack. An examination and analysis of the moisture, both in the room and in the masonry, can identify possible problems in hidden wood constructions that require closer examination (Fig. 5.17).

Wooden structures usually reach a moisture content of no more than 20%. If too much moisture is detected, there is probably a problem in the wall or roof. Rot, caused by fungi or insects, is not the only cause that needs to be identified. Excessive roof loads, changes in the use of the building with the resulting increased load on the

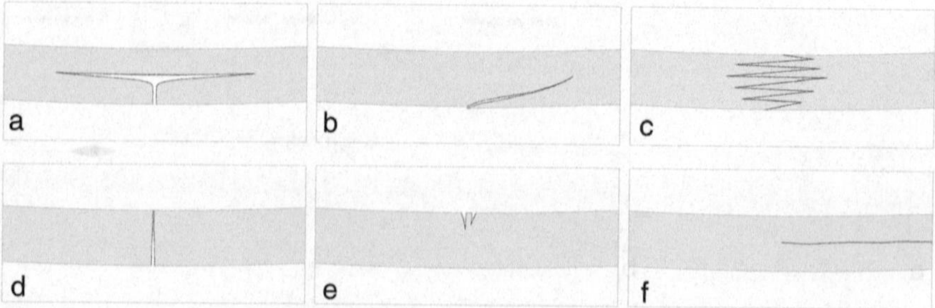

Fig. 5.17 Characteristic beam failure—**a** Simple tension, with stress parallel to the wood fibres. This is common when the wood is seasoned. **b** Transverse stress, with stress diagonal to the wood fibres. **c** Splinter stress, producing a large number of slight failures. This is often the case with hardwood. **d** Brittle stress, the fracture passes completely through the brittle wood beam. **e** Compression, common in green woods, where the fracture occurs at varying distances from the neutral plane of the beam. Often two or more fractures occur at about the same time. **f** Horizontal tension, shear fracture, where the top and bottom of the beam slide along each other

floor slabs, inadequate design criteria, improper material selection, and construction errors or lack of maintenance can all lead to decay in wood structures.

Cracks in wood structures are common in both old and new buildings. Although in most cases these defects are not a problem, they can sometimes be a weak point.

Cracking due to mechanical stress—A timber structure can fail due to an inappropriate configuration (geometry, size and connection) combined with static and dynamic load effects.

When a structure is subjected to excessive loading, it will crack or collapse in a characteristic manner. These cracks are classified as tensile, compression and horizontal shear cracks according to their origin and differ based on the fibrous nature of the wood. The following figure shows typical causes from which a beam can crack and many types of cracks.

Cracking due to the hygroscopic behaviour of wood—According to Franke, the moisture absorption caused by the hygroscopic behaviour of wood with shrinkage and swelling of the fibres determines the changes in the physical and mechanical properties of wood, such as strength values or modulus of elasticity. The changes differ according to the three axes of the material. A gradual increase or decrease in moisture content leads to dimensional changes in the cross-section and also to internal stresses (Figs. 5.18 and 5.19).

Joint failure—Causes of joint loosening include hygroscopic behaviour of wood, corrosion of metal fasteners, adhesive ageing, construction defects, and wood degradation. The resulting reduced bearing area creates eccentric stresses in the structure and subsequently fracture. The corrosion of the iron nails causes the wooden holes to expand due to the increase in volume of the metal. Then, in a second phase, the nails become thinner and can move in the holes, with the result that they are ejected.

Fungal decay often occurs between metal joints and wood or between beams and exterior walls, due to moisture and lack of ventilation.

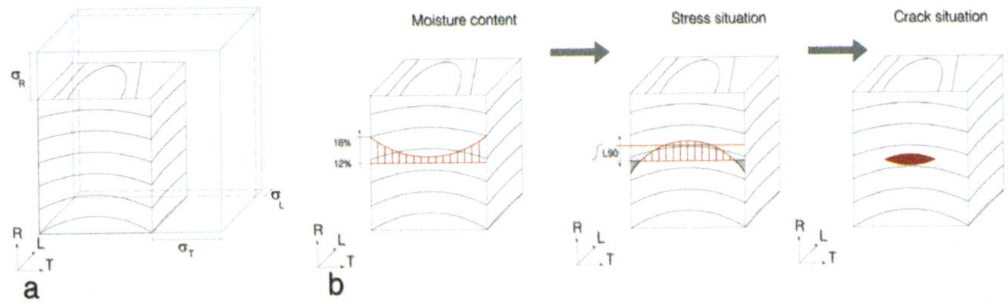

Fig. 5.18 **a** Varying shrinkage or swelling according to the axes of the material. **b** Gradual increase in moisture content and stress reaction with crack growth

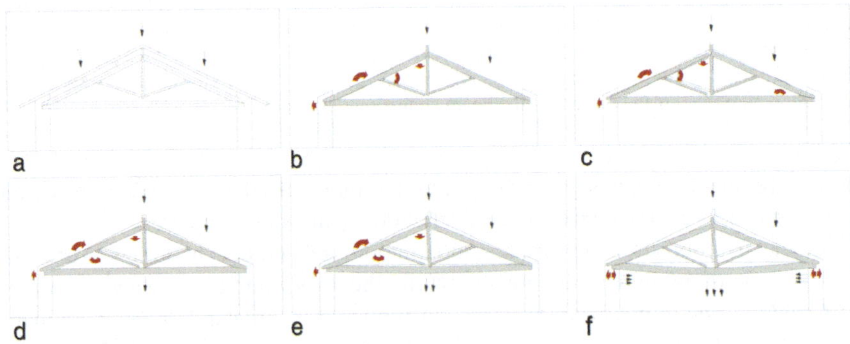

Fig. 5.19 Schematic representation of a truss fracture by separation. The final phase is the rupture of the tension member and the consequent collapse of the unit

Beams that can move due to decay of joints are subject to a combination of rotation and translation that can stress and fracture the structure.

Timber Decay

Moisture is one of the main causes of rot in wooden elements. Due to the attack of fungi or insects, wood decay develops at a certain humidity and temperature. Decay begins with gradual weakening and decomposition of wooden structures. The most important way to prevent insect and fungal decay is to keep the wood dry. For a correct diagnosis and the prescription of specific treatments, a detailed examination by specialists is necessary.

Fungal attack occurs only in a sufficiently moist environment and wood-destroying fungi belong to two families: *Serpura lachrymans* or *Coniophora puteana*. Both fungi, known as “dry rot” or “wet rot”, require more than 20% moisture in the wood. The fungi also differ in their optimum temperature requirements, but for most of them the temperature range is 20–30 °C.

Rotten wood fungi are often referred to as “brown rot” or “white rot”. The first fungus eats only light cellulose, turning the infested wood dark, while the second fungus feeds on both cellulose and lignin, turning the infested wood white. It is important to identify the fungus correctly, as the treatments are different.

Dry rot (Serpura lachrymans)—This fungus grows on wood that has remained damp for a long time, typically due to dampness in the building, often combined with lack of ventilation. Dry rot itself can produce the moisture necessary for continued growth. The optimum temperature for dry rot growth is between 23 and 25 °C; above 40 °C the fungus is quickly killed. The most vulnerable parts of the building are wooden roof structures and any joists that come into contact with the exterior walls. Thin, airborne spores mark the first infestation, which turns rusty red and without treatment gradually spreads as a white or grey coating. The fungus feeds on the wood and reduces its load-bearing capacity. The infested wood can be recognised by its light brown surface, which is structured by a combination of deep longitudinal and transverse cracks.

Wet rot (Coniophora puteana)—This second widespread wood-destroying fungus must develop a lot of moisture compared to dry rot. In an advanced stage, this fungus forms thin, dark brown strands and there is no sign of the first spores. The cracks are less deep than those of dry rot.

Insect infestation—Typical small circular or oval holes on a wood surface are signs of insect infestation. If no action is taken, there will be a deposit of wood dust and a gradual reduction in the strength of the wood structure. Also, when insects attack dry wood, moisture and fungus increase the vulnerability of the elements. It may take a long time for the process to develop. After a period of one or more years, larvae born from eggs laid by the insects in wood cracks grow and emerge from the wood surface, leaving holes characteristic of various insect species. The exit holes are generally clear and accompanied by wood dust.

Methods of inspection—Careful examination is necessary in cases of fungal infestation. The first objective is to locate dry rot, which may occur without visible symptoms. The goal of any investigation is to identify infested or at-risk areas and create a logical risk assessment. According to Watt (2007), two phases of investigation are necessary:

Step 1—Initial inspection:

- to detect typical features of decay, such as red, dust-like spores,
- to detect internal and external wall and roof moisture, actual or potential sources of moisture,
- analysis of cracks or defects in the wood structure,
- measurement of air temperature and humidity level.

Step 2—Once the areas of risk have been identified, further special investigations may be required. A superficial wall moisture survey may indicate deterioration in a hidden part of a timber construction.

Destructive methods—The two main methods of detection are: sampling with a drill to determine degradation and measuring moisture content.

Testing wood density and moisture content using a "kiln" are two tests that are commonly used. A very small amount of wood is required. The cylinder removed is weighed, then dried and weighed again to determine the water content.

Using wood screws is a quick and inexpensive system, although more sophisticated methods are available. Resistograph drills use a very fine micro-tip, the frequency of which can be recorded electronically.

Non-destructive methods—Non-destructive testing includes fibre optic testing, ultrasound, and infrared techniques. A fibre optic endoscope indicates internal degradation in wood structures; an ultrasonic technique is useful to determine the value of the elastic modulus of wood; a pilodyn can determine the density of wood.

Flat Roof Decay

Deterioration is related to a lack of routine and proper maintenance, errors in detailed design, and changes in the intended use of the building. Leakage and dampness are the most common problems with flat roofs. The causes can be many, but roofing details and membrane deterioration represent the highest percentage of causes for infiltration.

A careful examination of all details is required to identify any defects or movement that may have allowed rain to infiltrate under the waterproofing layer.

A poorly attached membrane can become dislodged by wind, and uplift resistance is drastically reduced if overlaps and flashings are not done correctly. Improper installation of the roofing membrane can cause shrinkage, surface erosion, blistering and cracking.

Improper roof slope or water drainage blockage on lightweight flat roofs will result in rainwater accumulation, which can lead to deformation due to weight gain. Precautions must be taken to prevent this phenomenon, such as appropriate overflow systems.

Condensation can reduce thermal insulation performance, cause damp patches, mould growth on internal surfaces and material damage. In uninsulated and unventilated flat roofs, moist warm air generated in the spaces below can become trapped under the waterproof covering. A vapour barrier must be installed on the ceiling.

Gap condensation can cause hidden damage; if this cyclic event occurs over a long period of time, large amounts of moisture can be deposited. This phenomenon can be easily detected by infrared inspections.

Due to the orientation of the roof, the materials are usually exposed to incident solar radiation for long periods of time. Although the roof radiates much of the energy back into the atmosphere, excessive temperature in the summer causes thermal movement and stress to the materials. There is a significant rise in temperature with a possible drop in performance. Each type of roof membrane may require specific protection systems.

5.3 Repair Techniques

As mentioned earlier, the most common cause of damage to buildings is moisture. Moisture sources can enter the structure from the exterior and interior, and a proper investigation is required to determine the cause of the moisture before any remedial action can be taken.

Old buildings with brick or stone exterior walls and massive ground floors are susceptible to rainwater infiltration and rising damp from the ground because waterproofing measures were not provided in the past. Underground walls are permanently vulnerable to groundwater penetration and it is very likely that these old walls had no waterproofing treatment. In such cases, their interior surfaces and finishes are exposed to moisture and decay.

Condensation is another cause of degradation. Moisture caused by condensation is misdiagnosed as water infiltration. There are several methods to reduce or treat the problem. The origins vary and are often caused by indoor activities that produce varying amounts of water vapour and cause surface or interstitial condensation. Damage caused by crack condensation can be more dangerous as it is not visible and leads to a reduction in the performance of the insulation layers.

Rainwater infiltration through the roof causes damage and deterioration of woodwork, insulation, ceilings, and interior finishes, as well as mould on the top of walls.

Most old buildings contain many wooden components and it is not uncommon to find degraded timber, especially if the building has suffered from neglect and lack of maintenance. Dampness is one of the main causes of wood decay, and fungal or insect attack leads to progressive weakening of wood structures and, if no action is taken to repair them, to their eventual decay.

When structural members have architectural value that must be preserved, the use of traditional materials, techniques, and finish details is generally preferable to maintain the integrity of the original details.

However, it should be remembered that traditional repairs may have limited structural performance and their improvement is often necessary. The use of contemporary materials, innovative techniques and details should always respect the historic value of the original timber structure.

For a number of reasons, it may be necessary to reinforce existing walls during renovation work. Conversion and renovation of existing buildings with major modifications often require careful assessment of the building system.

New construction or modification of existing construction with reinforcements and underpinning to increase the bearing capacity of the foundations must be well designed according to the characteristics of the building.

The close relationship between the building system, historic period, and deterioration must be carefully understood in order to plan appropriate repairs. Boxes 5.3.1 and 5.3.2 illustrate some possible repair techniques.

5.3.1 Box: External Wall Repair

Various techniques can be used to solve problems with the moisture penetration of solid walls.

Prevent Rising Damp
Rising damp in walls can result from these main causes:

- gap in a damp-proof course in the original construction,
- deterioration of the existing damp-proof course due to age or external causes,
- transition and overlap of the existing damp-proof course.

Buildings built before the 1940s do not have a damp-proof course, so many of them suffer from severe rising damp. Sometimes it is not possible to radically solve the problem, only to reduce it. Before deciding on the most appropriate remedy, it is important to define the best level to intercept the phenomenon (this means that the lower part of the wall remains damp).

For solid walls, three main criteria are usually applied. The choice depends on the characteristics (brick, stone or concrete blocks), the thickness of the wall, the coating and the decorations (Fig. 5.20).

Chemical barrier—A series of holes is drilled into the wall, just above the defined 'boundary' for rising damp, to form an impermeable barrier (usually about 20 cm from the soil). Devices inserted into the holes apply sufficient pressure to release silicone resin-based water repellents or aluminium stearate-based polymeric water repellents. The pressure facilitates water repellency from the pores and allows the resin to penetrate. Double-sided measures are recommended when the wall thickness exceeds 45 cm. Hydrophobing agents work on the principle that they line the pores within the material being treated, rather than clogging them.

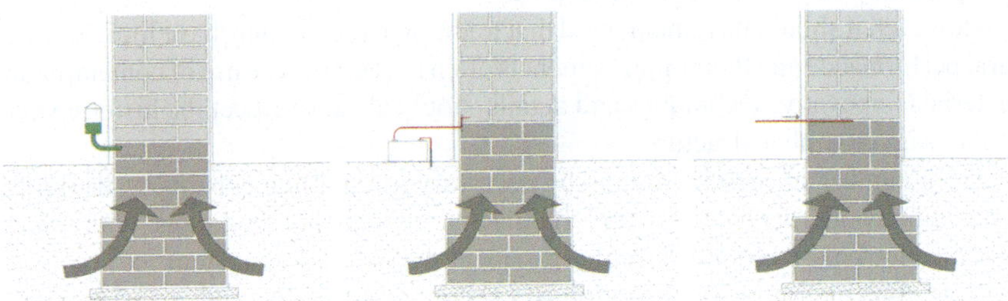

a b c

Fig. 5.20 Chemical barrier, main systems: **a** Pressure injection of chemical hydrophobic products.—Slow diffusion injection of chemical hydrophobic products. **b** Dehumidification, main systems:—Active electro-osmotic system.—Passive electro-osmotic system.—Electro-osmotic phoresis. **c** Mechanical cutting, main systems:—Insertion of a moisture-proof layer.—Insertion of a dense polyethylene membrane insertion.—Insertion of a stainless-steel foil

Due to the chemical stability of the resin and the presence of salts, the chemical barrier system must be repeated after a certain period of time. Inspections must be carried out to confirm that the injected liquid completely fills the capillaries.

Mechanical cutting—Horizontal cuts made at a rate of one metre per unit time and using a variety of techniques to allow the insertion of membranes and cement. This physical barrier determines a clear separation between the still wet wall below the cut and the upper dry part. This system is delicate and not always advisable, because if not well executed, it can lead to structural damage. This method can only be used on walls with continuous horizontal mortar joints.

Dehumidification—This technology is based on the principle that moisture rises through capillary action and the electrical potential created between the wall and floor. A small electric current through a series of copper or titanium anodes inserted into the masonry repels the rising moisture molecules down from the walls and back into the soil. This method is particularly suitable for high quality walls with frescoes that need to be preserved. Positive results depend on many factors, such as the intensity of the electric field, condensation that can occur inside or outside, temperature, capillary structure, wall thickness and the amount of moisture.

Prevent Moisture Penetration from the Soil

Many methods are possible, depending on the various repair options. Whenever possible, it is better to intervene on the outside of the wall. There are two main criteria (Fig. 5.21).

Barrier and drainage—Walls that are in direct contact with the soil are subject to moisture intrusion. The best method is to catch the water and place a waterproofing layer between the masonry and the soil. To keep the masonry dry, drainage systems that collect the water and drain it elsewhere are recommended, as well as wall ventilation.

Dehumidification—A new internally ventilated wall and raised floor can be used to separate rooms from outside moisture. When the system is vented, a layer of

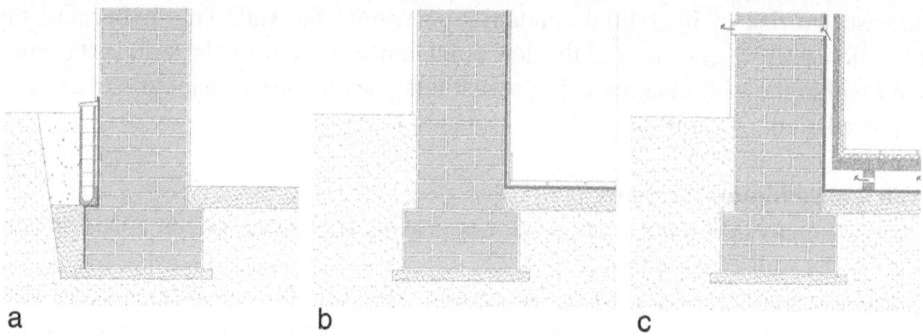

Fig. 5.21 **a** External repair, main systems:—Drainage ditches.—Ditches with water collection and aeration. **b** Internal repair, main systems:—Internal waterproofing on walls and floors.—Internal wall, with closed cavity.—Internal wall, with ventilated cavity. **c** Internal wall and raised floor, with ventilated cavity

insulation is expected to control thermal comfort. This system is usually used when it is not possible to operate the system from the outside.

Prevent Damp Penetration

For interior treatment, a variety of materials (cement, bitumen, or urethane-based) are available specifically for painting or spraying on walls with moisture above and below the floor.

Water-repellent solutions are available for exterior treatments, providing a simple and cost-effective way to protect walls from rainwater penetration, provided the problem is not too severe.

Exterior render finishes can be considered where other upgrading is required, for example to change the appearance of the building or improve thermal performance.

Prevent Condensation

Various measures must be taken to ensure that condensation does not occur:

- retrofitting or replacement of existing elements must not increase the risk of condensation,
- changes of use may lead to condensation, therefore appropriate measures must be taken to prevent the occurrence of condensation.

Three specific factors must be considered to control condensation phenomena: ventilation, heating and thermal insulation with a vapour barrier.

Natural ventilation helps to remove air containing water vapour from the building. In areas where large amounts of water vapour are generated, exhaust fans must be used. In highly thermally insulated buildings, extraction systems with heat recovery are required so that the temperature is not lowered.

If the wall surfaces of the rooms are kept above the dew point with suitable heating systems, surface condensation is prevented.

Thermal insulation reduces the risk of surface condensation by ensuring that the internal surface is maintained above the dew point. However, thermal insulation can increase the risk of interstitial condensation within the wall. This is because the insulation shifts the position of the dew point temperature from the wall surface to a point within the wall thickness. To prevent this phenomenon, a vapour barrier must be placed on the "warm side" of the insulation layer.

Wall Consolidation Systems

There are many different techniques for improving resistance, mortar adhesion, and altering the wall construction (Fig. 5.22).

Improvement of mechanical properties—To improve the mortar properties, consolidation mixtures are injected into the walls. These mixtures, filled into the existing joints, give continuity and compactness to the wall and make it more resistant. This system cannot be carried out in walls with abundant compact mortar, as the diffusion of the consolidation mixture is not guaranteed.

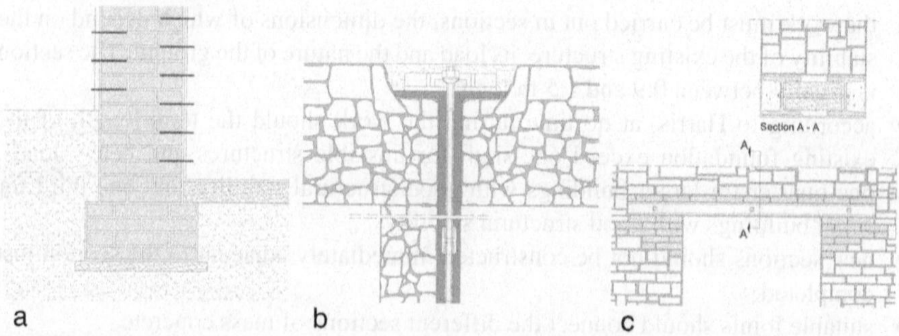

Fig. 5.22 **a** Improvement of mechanical properties, main systems:—Injection of compacted cement.—Resin injections. **b** Crack seams, main systems:—Armed stapling.—Metallic tie rods. **c** Opening, main systems:—Steel frame and grout injections.—Reinforced concrete frames and mortar injections

Crack seams—Steel bars are inserted into cracked walls with mortar or resin injections. Tie rods may be used with appropriate interlocking and joining systems to connect walls damaged by bending, rotation, or translation, or by bowing.

Opening—Two or more steel beams connected by tie rods are inserted into the consolidated bearing wall to create an opening. The number of beams depends on the thickness of the wall. Instead of steel beams, reinforced concrete beams are also suitable.

Underpinning Systems

A number of different reasons may require intervention. The most common are:

- excessive settlement and failure of the existing foundation, resulting in damage to the building,
- because of a new neighbouring building whose foundation is deeper than the existing one,
- because of an additional floor to increase the bearing capacity of the existing foundation.

Underpinning systems always involve excavation under existing foundations to construct a new building. This requires precautions because of the stability of the existing structure:

- existing loads should be temporarily reduced, if possible,
- additional work should be done to stabilise an existing weak structure, such as temporary bank structures,
- continuous monitoring should control any movement of the existing walls during the underpinning work using calibrated indicator lights, heavy duty plugs or laser monitoring equipment.

Mass underpinning with concrete is one of the most commonly used methods. This system requires that:

- the work must be carried out in sections, the dimensions of which depend on the stability of the existing structure, its load and the nature of the ground. The section is usually between 0.9 and 1.5 m long,
- according to Harris, at no time during the work should the total length of the existing foundation exceed one-sixth for unstable structures and heavy loads, one-quarter for larger buildings with good structural stability, and one-third for small buildings with good structural stability,
- new sections should not be constructed immediately adjacent to the section just completed,
- suitable joints should connect the different sections of mass concrete,
- expansion mortar fills the void (a few millimetres) left between the concrete pour and the existing foundation to provide the required resistance.

Concrete and Cement-Based Conglomerate Restoration

Preservation of historic concrete is a problem of appearance, as all techniques are visible and a top coat is often required. Elastomeric sealants are used to repair cracks and joints to reduce moisture and air infiltration. When used to repair cracks in historic facades, they can be visually disruptive.

Repairs—Many historic concrete facades show signs of deterioration due to flaws in the original construction. Any improvement must be thoroughly evaluated for compatibility with the original appearance, with many aspects to consider, including surface preparation, formwork installation, and concrete mix design.

In any concrete repair project, the process of investigation, laboratory analysis, and complete repair allows for both correct work and the implementation of durability measures.

Preparation for local repairs requires removal of the damaged concrete to determine the repair tent and find a solid base. The area must be cleaned, usually by careful sand and air blasting. Additional steel may be required if the existing rebar is severely damaged. Exposed existing reinforcing steel and other embedded steel elements should be cleaned, primed and given an anti-corrosion coating.

The repair material should be reinforced and mechanically bonded to the existing concrete. Reinforcing materials used in repairs usually include structural steel, epoxy-coated steel, or stainless steel, depending on existing conditions.

Special formwork is required to replicate decorative concrete details and to achieve special finishes and textures. Selection of the proper repair materials requires an evaluation of compatibility, performance, and properties and may involve laboratory testing of the proposed materials. A formulation should use aggregates similar to those in the original concrete mix.

Surface consolidation—Consolidation products, such as synthetic resins or inorganic resins, are applied until they penetrate porous areas and reach a suitable depth to prevent crusting and subsequent delamination.

Realkalization—This technique is used to restore the alkalinity of carbonated concrete. The concrete is impregnated with an alkaline solution, which in some cases is injected by passing a direct current through the thickness to the steel-reinforced level. These measures increase the alkalinity of the materials around the reinforcement.

Replacement—Ornamental elements can be prefabricated or cast in place to match historic elements and replace missing elements. Careful mix formulation, placement, and finishing are required. Trial repairs are used to evaluate proposed replacements and refine construction techniques.

Protection Systems—Film-forming coatings are often inappropriate because they could change the colour and appearance of a surface, and higher-quality coatings could alter architectural finishes and ornamental details.

Clear, penetrating sealers provide protection for lower-quality historic concrete and avoid major future repairs or replacement.

Cleaning Methods

There are several methods that must be selected depending on the characteristics, surfaces and complexity of the details.

Water cleaning—The surface is sprayed with low pressure water or atomised water. This work takes varying lengths of time, depending on the intensity of the contamination and the nature of the surface. Generally, it is necessary to combine this action with local cleaning with brushes to remove any remaining deposits. Excessive amounts of water and excessive pressure must be avoided to prevent erosion.

Chemical cleaning—Different substances are used according to the method. The duration of the process varies (from a few hours to a month), depending on the type and thickness of the contamination and the method used. The surface is then washed off with water. Adsorbent clays or biological substances have a significant effect against harmful layers of dirt. This technique is slow and limited to small surfaces and decorative details, and sometimes it must be repeated many times.

Micro-sandblasting—The surface is cleaned with a jet of lightweight abrasive (usually small diameter aluminium or glass balls). A device precisely directs the jet against the dirt at a variable and controlled pressure (from 0 to 6 atm.). This technique must not be used on large surfaces.

Laser cleaning—Black deposits are removed by laser beams that cause high temperature and vaporisation of the dirt. This system works selectively without damaging the surface. In fact, once the laser has removed the dark material, it stops because of the colour change.

Frame Consolidation

The replacement of damaged parts or the addition of elements that have gaps by the insertion of a "prosthesis" requires a careful study of the material properties and surface characteristics in order to reproduce them correctly and to select suitable adhesives or special mortars.

Surface consolidation—Consolidation products are applied to stone elements so that they can penetrate their porous structure. Synthetic resins or inorganic chemicals are most commonly used. Consolidants must penetrate deep into the stone to prevent crust formation and its subsequent detachment.

Deep consolidation—In areas where a layer of plaster is detached from the substrate, the injection of a consolidation mixture is practised. This mixture, which varies according to the size of the detachment, must be characterised by good penetration, not excessive mechanical force, porosity, and absence of soluble salts. Typically, an air–lime mixture diluted with a percentage of primer and possibly mixed with sodium gluconate or brick dust is used. The limit of this method can be a consolidation in the "leopard spot", without precise geography of the glued areas.

Filling—Deep and wide cracks are filled with a mixture of binder (organic or inorganic) and inert powder, which may be sand, silica, or the same stone material. Hydraulic binders or gypsum mixtures are not acceptable as they cause sodium or potassium salt efflorescence on contact with water and are time-limited.

Bonding—Pieces of stone and an adhesive substance are inserted into the original block with small spaces between them. Typically, epoxy liquid resins without inert additives form the adhesive. This epoxy should be used with caution as it is not reversible. For larger gaps, new, appropriately shaped pieces of stone are bonded to the original block with titanium or stainless-steel pins.

Protection—Protecting natural or artificial stone requires great care. Some solutions, even if protective, alter the existing surface, such as turning opaque into shiny, creating films that are altered over time by exposure to sunlight.

5.3.2 Box: Roof and Floor Repair

Special intervention techniques, such as resin fixing of steel or carbon fibre panels to existing timber components, require specialist knowledge for their design and specification.

Treatments

If the fungal or insect attack on the wood has not progressed to the point of significant structural strength loss, special nozzles are inserted into the wood to provide deeper pressure and better repair than surface treatments. The same treatment can also be applied to sound wood that is considered susceptible.

Dry rot fungus—Because of the characteristics of dry rot (it is widespread and capable of attacking dry wood), its removal affects not only the infested wood and its replacement, but also all adjacent elements and materials. An examination may be necessary to identify infested areas. The experts follow certain procedures according to the regulations.

Wet rot fungus—Because wet rot is less productive than dry rot, the rot is more localised and treatments are simpler. Typically, it consists of replacing the infested wood structure with a new, pre-treated piece, which also eliminates the original causes of the fungal infestation. However, if the infestation is widespread, local repair is not appropriate and complete replacement is preferable.

Insect—If the insect infestation is not advanced, it is appropriate to proceed with water-based insecticide treatments. If the infestation is inside the wood, the insect infestation must be well assessed to ensure that the organic solvent treatment penetrates to the proper degree. To prevent future infestations, pastes or injected solvent-based formulations should be applied to all wood structures.

Mechanical Timber Repair

Roof and floor structures are the most susceptible to moisture and water intrusion; roof trusses and joist ends inserted in load-bearing exterior walls are particularly subject to rot. A thorough examination of hidden parts is necessary to detect effective decay and program immediate repair before the structure collapses.

Eliminating moisture from load-bearing walls is not as easy, but it is possible to create a 5 cm gap between the wood and the wall and apply a water-repellent coating to prevent contact between the wet wall and the wood. Among the most traditional techniques for repairing wooden structures and improving their mechanical performance is the introduction of steel angles, often in combination with new wooden parts (Fig. 5.23).

If the deterioration is confined to the ends of the joists or trusses inserted in the wall, only these parts are cut off and replaced, and steel supports may be used to connect them mechanically with the wall and a concrete pour.

If there is some residual strength at the end of the truss, an alternative, depending on the severity of the damage, is to treat the damaged portions with preservative and install steel reinforcing plates on each side. These plates must be extended beyond the damaged part and overlap the sound timber by at least 150 mm. Direct contact between timber and steel must be avoided as condensation can form and promote rot or fungal growth.

An alternative to mechanical methods is the use of epoxy resin-based methods. The bonding capacity of the resin may need to be supplemented with fibreglass reinforcing rods or steel reinforcing rods and plates to ensure a successful repair, taking into account the degree of wood damage. These reinforcements, if used, must extend to the depth of sound wood to ensure effective bonding and load transfer, and the epoxy must be sufficiently fluid to ensure proper impregnation of the fibres. This technique requires highly specialised skills and is typically used when historic wood structures require special restoration measures (Fig. 5.24).

Resin-based repair methods are often used as an alternative to more traditional systems, for example, when a damaged piece needs to be removed and replaced with a new one. In this case, an epoxy resin part is bonded to sound wood using reinforced polyester rods embedded in epoxy resin.

If the existing wood structure is severely damaged, the entire member is removed and replaced. A new timber can be formed and joined with a tenon inserted into a pre-formed recess in the existing structure. Epoxy resin can provide good joints and fill in any gaps.

Stiffening with Steel or Wood

Floor slab structures are in good condition, but often need to be adapted and retrofitted to meet current building codes. Therefore, structural and seismic adjustments are always necessary. The original adjustment of the floor slab thickness leads to further

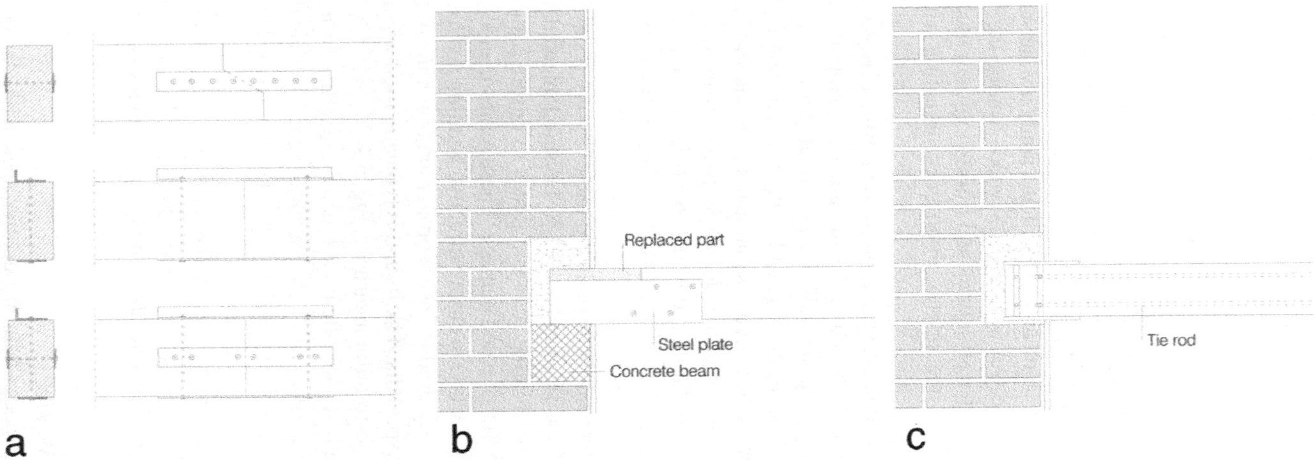

Fig. 5.23 Beam decay. Sometimes wall reinforcement is required. **a** Steel reinforcement. **b** Concrete beam and new section connected to old beam by steel plate. **c** Steel reinforcement and tension bar

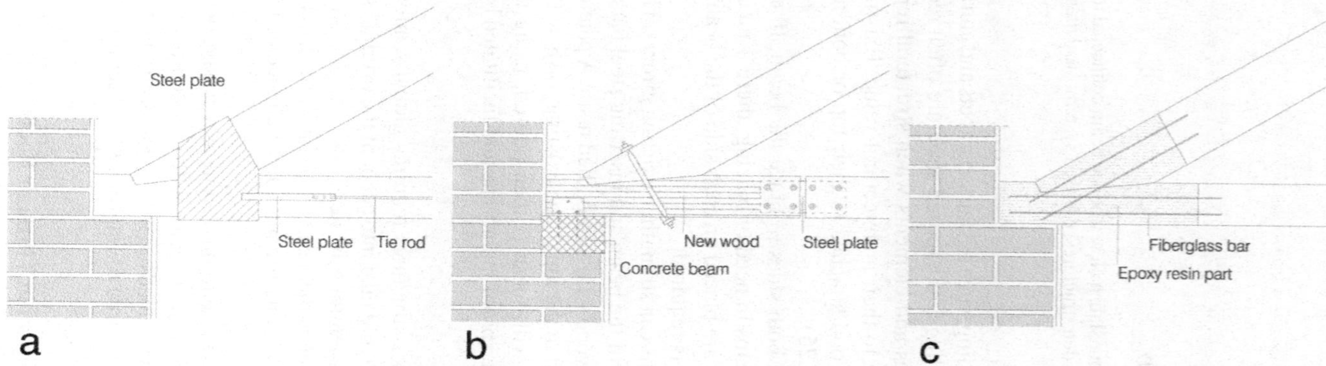

Fig. 5.24 Disintegration of the binding end. **a** Steel plates and tie rods. **b** New wooden part connected to the old beam by steel plates. **c** Epoxy resin part connected to the old structure by fiberglass rods

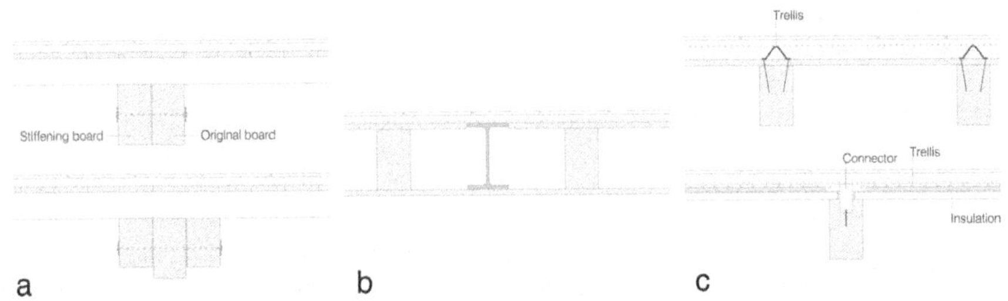

Fig. 5.25 Reinforcement of the floor slab. **a** New wooden beam connected to the old beam. **b** Steel beam inserted between the old wooden structures. **c** Mixed system: steel mesh connected to the old beam and common concrete pour

work on other parts of the building, such as stairs, parapets and entrances. Preservation of original floor surfaces, ceilings, or exposed structure often requires disassembly and reconstruction of surfaces and elements when structural repairs or integration of equipment is required due to their architectural value. Existing floor joists may be reinforced with steel or wood attached to their top or bottom surfaces. Various methods are available (Fig. 5.25).

Steel plates are placed on both sides to brace the beam. If secondary beams or rafters can be placed over beams in need of bracing, there is an economical method of repair. If bracing members are placed on each side of the beam, a more expensive and time-consuming method is required.

Placing steel reinforcements can strengthen existing girders. This method involves cutting off the centre portion of the beam and inserting steel bars that are embedded and bonded into the existing wood with thin-bodied resin. A patented system consists of a steel truss that is attached to the top of the existing joists with mounting screws. A concrete layer reinforced with mesh is then poured, using the original plan as formwork, protected by a waterproof membrane and facilitated by insulating panels.

Global Reinforcement

Sometimes a system of timber structures requires global reinforcement to connect the elements together. This should take into account the original structural behaviour and any effects of the new structural scheme.

External steel bars can connect the beams of a slab or the truss of a roof to increase the stability or load-bearing capacity. Beams can be connected to load-bearing walls with tie rods.

Steel connections in a truss system can operate as a passive or active structure with a slight pretension. This reinforcement helps with the occurrence of high stresses and also absorbs some of the service loads to reduce the deformability of trusses. Reinforcement systems may be required to improve the structural stability of a pitched roof (Fig. 5.26).

Preventing Rainwater Penetration Through Roofs

Rainwater penetration through a roof not only causes inconvenience to the occupants, but often leads to deterioration of the structure, walls, insulation and interior

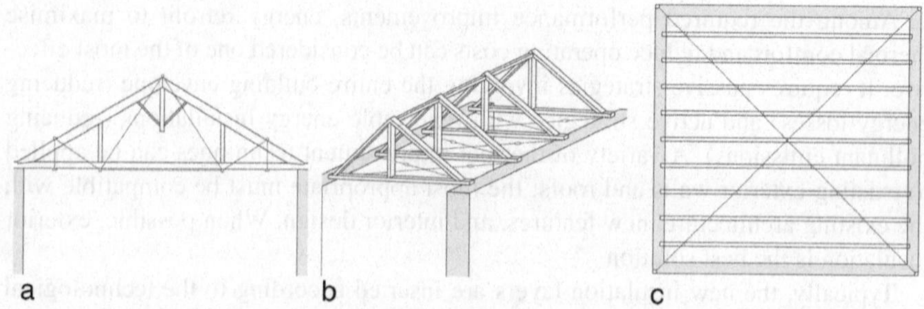

Fig. 5.26 Global reinforcement. **a** Reinforcement of the truss. **b** Connection between all truss girders. **c** Connection between beams and walls of a floor slab

finishes. To prevent damage, it is important to schedule seasonal inspections and take immediate corrective action when rainwater intrusion is evident.

A waterproof membrane can be installed between the roof tiles and the underlying layers in a pitched roof. In old buildings, the original roof tiles are often damaged by frost, chemical or mechanical damage and cannot prevent rainwater penetration. In this case, a new coating is required.

Flat roofs are more susceptible to rainwater penetration due to the low drainage rate and some technical details. Parapets, flashings, rainwater openings and joints between membranes are typically susceptible to degradation.

Waterproofing membranes have a service life of 10–30 years, depending on the material used, and are constantly exposed to the elements and mechanical stresses. In the event of a stormwater intrusion, a thorough investigation is required to determine where the water intrusion is located in order to correct the problem. Sometimes an instrumental investigation is required to trace the water path.

Original defects and work done during the life of a building are often responsible for the deterioration of the flat roof. In such cases, it is necessary to redesign these parts to avoid possible future deterioration.

To maintain a good level of performance, a maintenance process must be well planned; many complex and delicate details must be checked each time an external event occurs.

5.4 Improving Building Performance

In a renovation project, it is first necessary to carefully analyse the current building codes. The important evaluations concern the level of performance in terms of energy, comfort, safety and accessibility.

The change of use brings new floor plans that must be examined from many different angles. The overlap between the new and the existing requires careful evaluation of technical and economic feasibility.

Among the required performance improvements, energy retrofit to maximise thermal comfort and reduce operating costs can be considered one of the most effective. It requires passive strategies involving the entire building envelope (reducing energy losses) and active strategies with renewable energy installations (reducing pollutant emissions). A variety of thermal improvement techniques can be applied to existing exterior walls and roofs; the most appropriate must be compatible with the existing architecture, new features, and interior design. When possible, exterior insulation is the best solution.

Typically, the new insulation layers are inserted according to the technological characteristics of walls and roofs; calculations and software simulations are necessary to define the right solution, check thermal bridges and analyse the effect that the new layers will have on the overall performance of the building.

In addition to improving the energy performance of the building envelope, acoustic performance should also be considered. A wide range of techniques and solutions are available. For example, choosing high efficiency windows can provide good energy and acoustic performance.

The second important performance improvement relates to fire safety regulations. Some increase in fire resistance and changes in design for safety defined by new activities and number of occupants may be necessary. Passive strategies involving protection of steelwork, specific cladding materials and furniture, escape routes and escape staircases need to be complemented by specific active systems.

Old buildings, unless recently remodelled, do not meet accessibility standards for people with disabilities. Often the features of a historic, listed building make adaptation work difficult and require major alterations with additional lifts and stairs. This will be an important part of the architectural design that must make the new elements compatible with the existing historic features.

5.5 Conclusion

Identifying building technologies that will be used in a building in need of renovation is essential. There is often a lack of information, and only technical knowledge and proper research can reliably identify features and construction details that form the fundamental basis for subsequent analysis and design decisions.

The older the building, the more local investigation of the materials and techniques used is necessary. Reference to local period manuals will be helpful for a focused detection and investigation process.

Knowledge of degradation mechanisms and their correlation with specific construction techniques will allow selection of the best repair techniques. There are many methods available today for repairing buildings and improving their energy efficiency, which must be carefully selected and well evaluated according to their objective impact.

Chapter 6
Energy Retrofit of Suburban Districts: An Opportunity for Social Regeneration

Abstract Residential construction is the largest segment and accounts for 75% of the building stock in the EU-27, of which 36% is multi-family. It is generally accepted that the building sector is one of the largest energy consumers in Europe and residential buildings are responsible for the majority of energy consumption (more than 65% of energy is used for heating). A significant proportion of today's social housing in Europe was built between the 1950s and 1970s, when post-war reconstruction and heavy immigration flows to the most industrialised areas required an immediate response to housing shortages. In many cases, solutions were adopted that were very different from those of the first working-class neighbourhoods, which were built in the early 1900s primarily to address health emergencies and had deficiencies related to outdated building types and poor performance, lack of services, distance to work, and social degradation. This chapter demonstrates how the necessary energy refurbishment required by current regulations provides an opportunity to modernise housing to meet new needs and regenerate suburban areas. New housing models, small temporary apartments associated with multifunctional spaces, represent the latest requirements that characterise a new type of users and families. The recent emergency in Covid 19 has highlighted that this type of accommodation is not able to respond to ever-changing needs and requires flexibility to fulfil functions not strictly related to housing. The examples provided in this chapter are evidence of good practice in reviewing outdated models of existing housing stock.

Keywords Social housing · Inhabitants' needs · New residential models · Energy retrofit · Suburban regeneration

6.1 Introduction

The post-World War II metropolitan neighbourhoods built on theoretical models of self-sufficiency, which soon proved to be failures by becoming dormitory quarters, characterise many suburbs of industrialised European cities. Experimental, prefabricated building technologies, low-cost and geared to the desire for rapid construction, have produced standardised and anonymous housing types that today do not meet the

M. Grecchi, *Building Renovation*, PoliMI SpringerBriefs,
https://doi.org/10.1007/978-3-030-89836-6_6

needs and performance requirements of users. The recent Covid 19 emergency experience demonstrates the critical nature of these minimal accommodation and housing types and their inability to respond to ever-changing needs, creating issues around physical space and facilities for social distance, smart-working and e-learning.

A significant proportion of the housing stock in Europe today was built between the 1950s and 1970s, before any kind of energy legislation.

The housing estates built in the early twentieth century were a response to health and social emergencies, to the need for 'decent' housing and communal spaces, aimed at socialisation and mutual support between residents, activities to educate young people and keep them employed. Many of these were restored and no longer used by the poorer working class. Some considerations of building typology and neighbourhood morphology are discussed in Paragraph 3.

The deterioration in the quality of apartment buildings is causing the wealthiest families to move into higher quality apartments, while the other owners of the apartments do not even have sufficient funds for routine maintenance. As a result, the buildings often become dilapidated housing for low-income families. Consequently, the large-scale modernisation of the existing housing stock and the improvement of energy efficiency are important tasks for European societies.

The need to refurbish them to the level required by current legislation is an opportunity to rethink these buildings and make them more suitable for new housing requirements. The inadequacies do not only concern the buildings, but it is also necessary to rethink the collective spaces, the green areas, which, as in some emblematic projects developed in Europe, will lead to major changes. The introduction of a functional mix is also strategic to start a process that recognises attractions and points of interest even for those who do not live in the district.

New housing models, small temporary apartments associated with multifunctional spaces, are the latest requirements that characterise a new type of users and families. The recent Covid 19 pandemic has shown that housing needs to be flexible to accommodate a variety of activities and allow cohabitation that makes it a safe place for the family. The constraint of accommodating multiple family members in a confined space that is not equipped for smart-working and e-learning at the same time has led to criticism and discomfort during the lockdown period.

Connections to the city centre, places of interest and work (measured not by real distance but by travel time and efficient public transport) are other key issues that need to be addressed in order to rethink the periphery and make it an integral part of the city.

We will address two research questions: How can we meet the ever-changing user needs through the necessary energy renovation of existing residential areas? Can a functional and social mix be a correct and sufficient approach for the redevelopment of social housing?

6.2 Social Housing Neighborhoods: An Overview of the Current Situation

The decay of urban suburbs requires integrated interventions to redesign the existing building stock, especially the social housing built after World War II and until the end of the 1970s, when the adoption of thermal regulations began, taking into account energy problems. But we also have to evaluate the housing of the 80 s and 90 s based on the needs of the users and the quality of the urban spaces.

The obvious obsolescence of these buildings highlights the need for renovation, taking into account both functional and performance requirements. Improving the quality of space and living must go hand in hand with energy efficiency and resource conservation.

Passive and active measures should be part of an integrated planning strategy aimed at improving housing layouts and introducing solutions that can respond to ever-changing demands. Recent European experience shows that it is possible to combine functional, technological and energy retrofits in an integrated strategy that can lead to an overall improvement.

In Europe, the household sector accounted for 27.2% of final energy consumption in 2017 (Eurostat, 2019a). Looking at the targets set for 2030 to reduce energy consumption and CO_2 emissions, many measures still need to be implemented. In the EU-28, 22.3% of dwellings were built before 1946 and 44.1% in the period 1946–1980 (Eurostat, 2017a), which means that more than half of existing dwellings need energy efficiency improvements.

New and increasingly complex requirements arise from constant changes in the composition and age of the population, as well as factors related to work, expectations, models proposed by the media and the presence of diverse and extremely heterogeneous social communities. In 2018, 21.3% of urban dwellers in the EU-27 faced pollution, dirt and other environmental problems, and 12% of families with children reported little or very little satisfaction with their homes (Eurostat, 2020).

Housing quality is a broad definition, encompassing a variety of aspects that relate not only to the home itself, but also to the wider living environment in which people live. Noise, pollution and feeling unsafe in the neighbourhood are the most commonly cited factors by respondents.

Expectations are constantly changing, and meeting new needs inevitably means that the housing model needs to be redesigned and evolve to meet future demands. The most important data concern the ageing of the population, unemployment, single parents, severe disabilities, social hardship and immigration.

Understanding people's level of satisfaction is an extremely complex topic. From the studies of sociologists, it can be deduced that an important condition is the promotion of relationships through collective spaces and aggregation functions. With regard to the sense of belonging, it then becomes apparent that with the transition from the purely private to the collective and public sphere, the sense of space as part of one's living environment is lost and often becomes an object of vandalism and degradation.

Major neighbourhood redevelopments in Europe are beginning to redesign public and collective spaces. Here are some reflections on the phenomena affecting social housing neighbourhoods.

Family—This remains the most important reference for the housing project. However, due to social, numerical and immigration phenomena, the family is constantly evolving. In 2016, the two most common family types in the EU-28 were those consisting of one or two members. They each accounted for almost a third of the total number of households (Eurostat, 2017a, b).

According to the statistics, there are currently three important factors to consider:

- The decline in births and the widespread availability of post-war housing for large families. The gap between live births and deaths in the EU-28 has narrowed significantly since 1961, and in 2018 the number of deaths exceeded the number of births (Eurostat, 2019b).
- The number of single-parent families is increasing, leading to the need for support for their children during working hours. The 2011 Population and Housing Census shows that families consisting of single parents accounted for 16.0% of the total number of families in the EU-28.
- Growth of multi-ethnic people with different cultures and large families. Net migration in the EU-28 has increased significantly since the mid-1980s. The population increase recorded for the EU-28 in 2018 was therefore due to net migration. In 2018, net migration and statistical adjustment accounted for an increase of almost 1.5 million persons (Eurostat, 2019b) (Fig. 6.1).

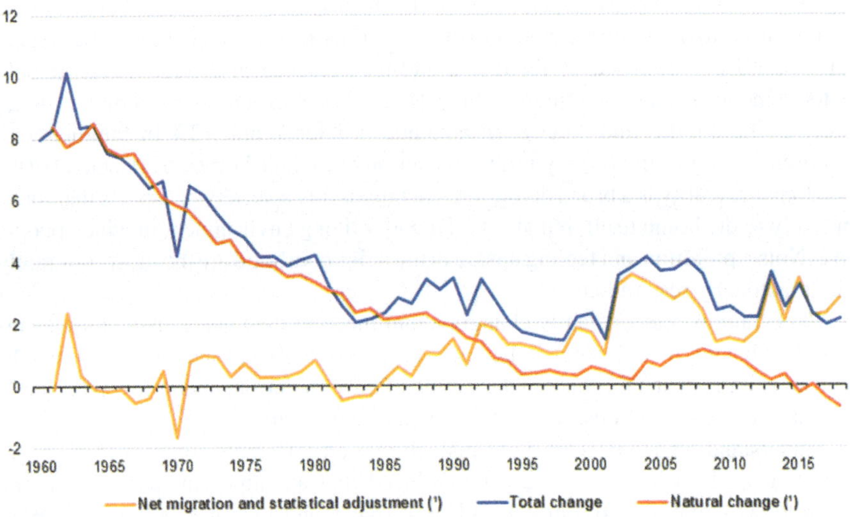

Fig. 6.1 Population change by component (per 1000 persons), EU-28, 1960–2018. (*) 1960 not available. *Source* Eurostat (online data code: demo_gind)

There is, however, an important contradiction: while it is true that the average family is characterised by one or two children, the increase in non-European immigration means a new presence of large families, which are now often forced to live in difficult conditions.

The second requires a reformulation of housing with autonomous but interconnected spaces and the need to design district structures dedicated to "baby parking" and activities for children of different ages to ensure support during out-of-school hours.

An ageing population—This is an established fact, but even in this case it is necessary to reflect on some phenomena that have opposite effects on life. The Population and Housing Census shows that in 2011 there were 20.6 million people aged 65–84 in the EU-28 alone. Looking at people aged 85 and over, 39% of the population lived in one-person households (Eurostat, 2017b).

Older people are affected by very different, often diametrically opposed, conditions. On the one hand, older people who are in good health can provide valuable support (e.g., grandparents helping with childcare) and who, having a lot of free time, need alternative activities and meeting places in their neighbourhood.

On the other hand, an important segment is characterised by older people who need help in various forms, leading to a rethinking of the mix of housing types aimed at different users who can establish mutual assistance relationships; monitoring rooms (carers for monitoring and assistance, remote care and medical care) and sheltered housing.

Unemployment—The widespread presence of the unemployed in any suburb inevitably leads to social hardship. In 2018, 6.9% of the young population of the EU-28 (aged 20–24) faced severe housing shortages, and 23.6% of the young population (aged 15–29) in the EU-28 lived in overcrowded housing (Eurostat, 2019c). In 2017, 113.0 million people, or 22.4% of the EU population, were at risk of poverty or social exclusion. This means that around one in five people in the EU experienced at least one of the following three forms of poverty: monetary poverty, severe material deprivation or very low household work intensity. Women and young people are most affected (Eurostat, 2019d).

Experimental solutions have been used to introduce new activities in specially created spaces (run by the partners who also guarantee their management) instead of "incubators" to support new work initiatives, or at least to set up a focal point to help young people in particular to find work and improve their level of education (Eurostat, 2019d).

Social discomfort: drugs, disability, and crime—Reality shows that social discomfort is exacerbated or generated by the degradation that characterises the periphery. The widespread presence of disabled people is not just a problem of adapting spaces and equipment; often the most extreme conditions result from serious psychological problems, and families often have to cope with uncontrollable situations. Following recent welfare reforms, dependent people such as the elderly, psychiatric patients and the homeless are now expected to live independently (Costarelli et al., 2019) (Fig. 6.2).

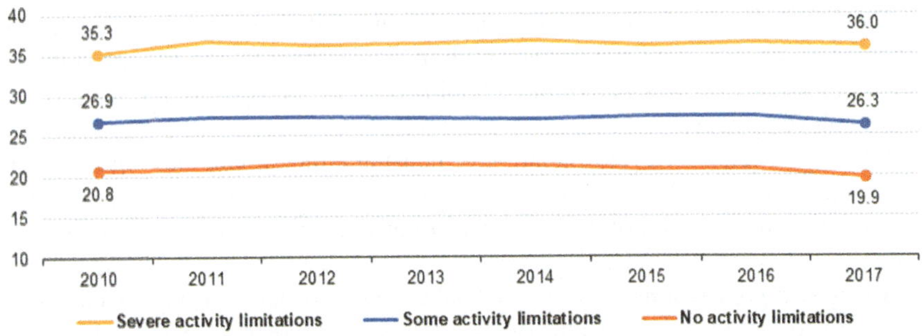

Fig. 6.2 People at risk of poverty or social exclusion by level of activity limitation, EU-28, 2010–2017 (% of population aged 16 or over). *Source* Eurostat (online data code: hlth_dpe010)

Another devastating phenomenon is the proliferation of drugs, which subsequently leads to crime, with an inevitable discomfort for the population, who perceive the problem in the form of fear and insecurity in some areas. The most common solutions used concern control devices, upgrading spaces perceived as dangerous, poorly lit and neglected. In 2018, slightly more than one in six people (17.4%) living in cities in the EU-27 Member States reported crime, violence or vandalism in their area (Eurostat, 2020).

6.3 Regeneration Process

Interventions in social districts are often carried out in emergencies, when postponement is no longer possible. Rather, it is necessary to plan a conscious regeneration of the existing stock, in some cases with the courage to demolish, to reduce density with the advantage of collective urban spaces. Resident participation is one of the pillars of the process: even if participation inevitably slows down project development, it ensures appropriate responses and the responsibility to maintain the acquired quality over time. Across Europe, there is a trend towards more user participation in service provision. User involvement measures are triggered by national policies and/or by proactive local initiatives. In terms of social housing, resident participation practices arise from a mix of provider organisational culture, regulation and relationships with tenants' or residents' unions and associations (CECODHAS, 2011).

Over the years, residents have expressed their needs in forms of self-adaptation of their accommodation (e.g. storage rooms, closing loggias with windows to increase living comfort and creating unplanned conservatories). These unplanned solutions are an inspiration for planners, as they show the different needs in the use of space.

Equipping neighbourhoods and buildings with flexible collective spaces and new unit models that can be regenerated according to changing needs can ensure continuous improvement in measures to support residents. Finding solutions to ever-changing phenomena means rethinking life as a series of different needs motivated by social emergencies, age groups, cultural traditions, work and educational activities, and unexpected situations, as the recent Covid 19 pandemic demonstrated. Then, of course, there are the necessary adaptations necessitated by new standards for energy conservation, fire safety, and accessibility.

Not only the redefinition of housing models able to follow the evolution of families, but also the introduction of "special" housing typologies designed for specific users (assisted senior housing, student housing, minimum units for temporary use, to name a few) can ensure a diversified presence of users, some of which are stable, but also constantly renewed. The debate on social mix now focuses more on ethnic than economic aspects and is perceived as the real problem of integration (Van Kempen & Bolt, 2009). As Costarelli et al. (2019) stated, social mixing can be seen as a dynamic and mobile accumulation of practices, "in recognition of the importance of personal attitudes and motivation to interact with neighbours from different socio-cultural backgrounds, beyond differentiation in terms of ownership and income levels".

Some projects in Europe have shown how it is possible to regenerate a neighbourhood, starting with the renovation of public spaces involving residents who discover that quality of life is not limited to the private space of homes.

Overcoming the monotony of many post-war neighbourhoods, whose vastness often does not allow for any dimension of social community and interpersonal exchange, is one of the most complex goals.

It is much simpler to redevelop the first public housing neighbourhoods built in the first fifty years of the twentieth century to solve health and migration emergencies. Many solutions are examples of typological experiments that were innovative for the time.

The morphology and size of the blocks (usually with courtyards) allow for an appropriate mix of private and collective accommodation. The size of the spaces allows for increased floor plans and typological diversification.

Figure 6.3 shows an example of a neighbourhood in Milan built in the early 1900s. It is one of the cases analysed in a study developed by the author for the ALER (Lombard Residential Building Company) to assess the tendency of public housing in Milan to renovate according to new housing models (Grecchi, 2008).

The strengths in the case of renovation can be summarised as follows:

- location no longer suburban and well developed,
- good architectural quality,
- clear hierarchy between outdoor spaces and gardens,
- layout concept that allows a good degree of flexibility (possibility to increase and/or reduce the number of entrances),
- spaces, excluding services, larger than the current minimum,
- optimal proportions between volumes and open spaces that define a pleasant urban environment.

Assuming that needs and expectations change over time, it is important to envision a renovation that includes measures for flexibility. They may concern plants and walls, the layout of rooms, the enlargement and/or division of toilets to ensure simultaneous

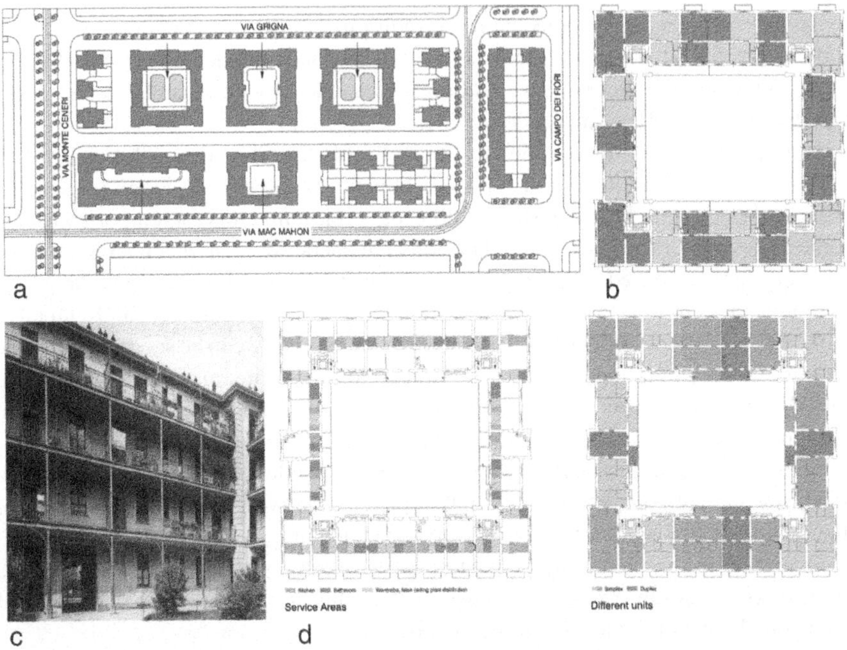

Fig. 6.3 Mac Mahon District built in 1908 by the City Council in the north-western suburbs of Milan (it was the first public social housing). The neighbourhood consists of four-storey buildings with enclosed courtyards distributed by balconies and two-storey townhouses with private gardens. The building types, while common and consolidated, have interesting innovations: the depth of the balcony (about two metres), which was intended as a place for meeting and social exchange; the variety of apartments (from studios to four rooms); and the introduction of toilets and running water in each unit. One problem was the lack of a service area, which of course does not meet the new expectations and standards. The service area (bathrooms, kitchens and cloakrooms) was placed in an intermediate position due to the great depth of the rooms. Plants are concentrated in a continuous false ceiling, made possible by the height of the apartments. **a** Masterplan. **b** Standard floor. The distribution is a mixture: through the balcony or directly through the stairs in the corners. **c** Historic photo of the courtyard (*source* ALER). **d** The hypothesis shows the tendency of this typology to be easily adapted to new needs. The strategic location of the duplex units allows the balcony to be divided and used for different purposes. It becomes a space that can be reinvented in the form of additional and flexible private spaces or common areas. *Source* The author

use, a layout that allows for variation of entrances. If a dwelling is designed as a static object, it is destined to become inadequate and obsolete (Scuderi, 2019).

A good degree of flexibility can be achieved by appropriate positioning of plants, toilet and kitchen blocks: An equipped, continuous and accessible cavity can ensure additions and changes over time. Proven examples of new builds can be adapted during renovations, with specific constraints being considered from time to time.

The majority of social housing, even if not recently modernised, has both vertical and horizontal plugs that do not meet current standards. In terms of size and accessibility, they are often cramped and poorly lit places that make those who traverse them feel uncomfortable. The adaptations required due to disability and fire regulations are also a way of reducing discomfort.

Main entrances connect urban and private spaces. They are often anonymous, small and of poor quality. Atriums and empty spaces on the ground floor (e.g. post-war "Pilotis" buildings) can be enriched with new functions, becoming an organised collective place. Distribution through galleries certainly allows maximum flexibility, but requires solutions with double exposure and respect for the necessary privacy.

A gallery can be a corridor separated from the façade, combined with duplex apartments and therefore located on alternative floors, with sufficient depth to accommodate other functions, both private and collective, or it can represent a space "available" for future changes.

6.3.1 Strategies for Retrofitting the Building Envelope

While demolition and reconstruction ensure safe homes with adequate levels of performance, in most cases the process takes an estimated 80 years to offset the use of natural resources and the waste generated (Scuderi, 2019).

Many approved research projects in Europe focus on the leading role of building envelopes in energy renovation and ignore other needs expressed by occupants.

Despite the various renovation projects developed recently, few propose fully integrated solutions. To maximise the benefits of often very expensive interventions, it is essential to consider the complex dimension of the problems and advocate an integrated and multi-faceted campaign of energy, structural, functional and architectural renovation.

The building envelope is an important "resource" in the global rehabilitation of the building, especially in cases where there are no historical constraints and no particular architectural value. In view of the 2050 energy and decarbonisation targets, it has become apparent that the envelope is becoming the most strategic source for retrofit (Konstantinou, 2014). One of the key differentiating factors in the choice of solution will be the presence or absence of people in the building during the works. Different retrofit options can be considered to improve the performance. They can be described and categorised as two-dimensional strategies (Konstantinou, 2014):

- ***Replacement***, with demolition and replacement of new facades. This is usually used when the deterioration of the hull is critical and repair requires too much work.
- ***Add-in***, with insulation layer inside. This is the case of façades that do not require repairs or have architectural value. The solution can be critical in the case of thermal bridges.
- ***Wrap-it***, when new and more efficient layers overlap with the existing envelope. Many different technical solutions are available, from simple protected insulation to double screens with ventilation. They allow interventions in the interior of the building (Fig. 6.4).

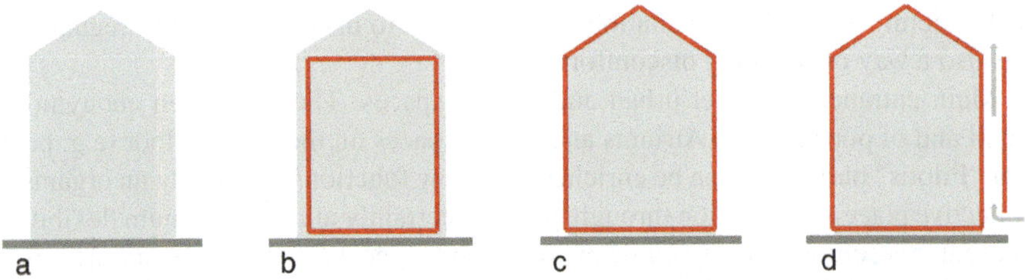

Fig. 6.4 Schematic example of building modernisation. **a** Existing building. **b** Modernisation of the inner building envelope. **c** Upgrading of the outer building envelope. **d** Upgrading of the outer envelope with south-facing double skin, shading and seasonal ventilation. *Source* The author

The complexity of the problems associated with the refurbishment of large social housing estates, built in prefabricated slab form after the Second World War and widely used throughout Europe, requires careful assessment and new approaches. Recent European experience shows that it is possible to combine functional, technological and energy retrofits thanks to an integrated planning strategy that takes into account the needs of the new users.

The patents for prefabricated panels used in the 1960s and 1970s are all characterised by minimal thicknesses, factory-assembled layer combinations, locally made joints with materials and technologies that were not suitable to be durable in the long term, especially in terms of panel size and weight. Their frequent use resulted in a succession of monotonous, repetitive and anonymous units and volumes, inflexible and rigid dwellings, and inadequate performance of facades and roofs that were often severely damaged.

The technologies and products available today represent an inexhaustible supply of possibilities; it is necessary to evaluate which are the best solutions able to optimise all the specific needs, not only in terms of comfort. Despite the numerous interventions in the building envelope caused by the European directives, many existing buildings still have deficiencies in terms of thermal performance: The need for an energy retrofit can become an opportunity to revitalise a building by improving its spaces, habitability, organisation and integration of installations. Lightweight constructions, achieved through the aggregation of pre-assembled industrial products based on environmentally sustainable and reversible approaches, are technologies that guarantee a faster management of the building site (sometimes occupants cannot be moved). As flexible models are necessary, technologies and facilities that can be easily modified, dismantled and/or implemented over time must be chosen.

The roof can also be a resource. Renovation can be designed together with other strategies, such as the reuse of attics, the introduction of new volumes, perhaps with variable and flexible goals that also allow temporary parking for occupants during the works (Grecchi et al., 2012).

Elevating a building requires not only careful engineering considerations, but more importantly urban-scale assessments. It is often necessary to reduce the density of a neighbourhood in order to create a better living space, sizing voids and gardens to define a balance between the built and urban environment. It can be assumed that

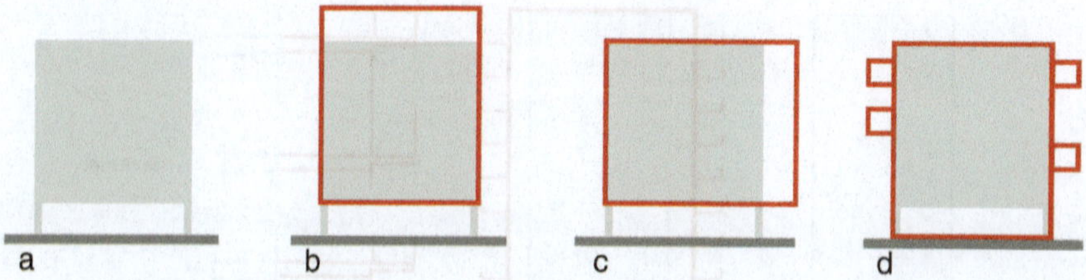

Fig. 6.5 Schematic example of building modernisation. **a** Existing building. **b** New volumes above the roof. They can be used as temporary accommodation for moving people during the works. **c** External volumes. They improve the available space of the apartments (e.g. a "buffer zone" to store energy and create different seasonal situations, see the examples of Lacaton & Vassal) or a new distribution system (elevator and fire escape) with external corridors that can increase the flexibility of the old floor plan. **d** New volumes to give more space to the apartments and common areas (e.g. use of "pilotis" on the ground floor). *Source* The author

a demolished volume will be redistributed to the preserved buildings. In this case, different strategies can be considered that introduce additional spaces in the facades rather than through the roof. Regardless of the strategy, two conditions are essential: the choice of light and "reversible" technologies and a careful architectural dialogue with the existing context.

Building renovation must meet the most immediate expectations of the occupants, taking into account the private habitability of the spaces. Envelopes are mainly considered to reduce energy losses, but how much can some "three-dimensional" solutions contribute to the renovation of interior spaces? And to what extent can new apartment floor plans optimise the "energy control" of the building?—simple concepts to illustrate the importance of optimisation solutions, taking into account different approaches and efficiency targets (Fig. 6.5).

Galleries, balconies, and overhangs become shading systems; "double skins" designed to contribute energy include loggias or horizontal pathways; continuous living spaces or double volumes can enhance natural ventilation; new layers added to the façade can embed plant cavities, for more flexible and cost-effective transformation over time. Three-dimensional solutions can simultaneously increase the performance of the building envelope and respond to user needs:

- ***Storey extension***, additional storey above existing roof.
- ***Additional volume***, to increase living spaces.
- ***Boxes connected to the envelope***, such as buffer zones, greenhouses or common areas.

"Transformation instead of demolition" is the guiding principle of architects Druot, Lacaton & Vassal for their study PLUS, a model defined in 2004 and applied to several cases of renovation of social housing. They wanted to counter the decision of the French government, which decided to demolish and rebuild most of the housing stock from the 1960s, using heavy prefabricated panels. The model combines energy efficiency and the strategy of passive construction with the space requirements, flexibility and needs of the users, wrapping the existing structure with large winter gardens

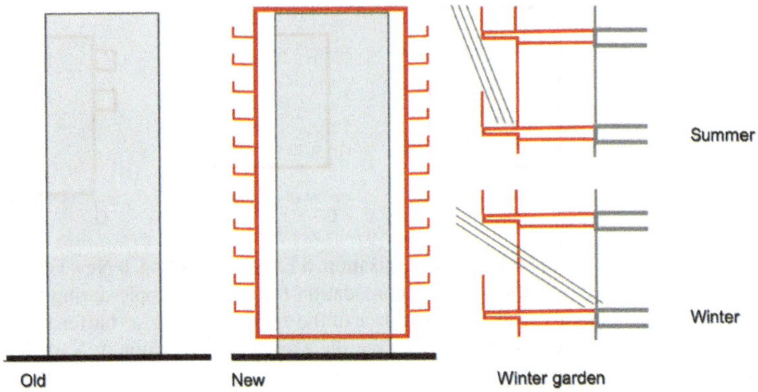

Fig. 6.6 Concept developed by Druot, Lacaton & Vassal Architects. Winter gardens increase the living space, provide effective thermal insulation in the cold season, and in summer, together with balconies, shade from the sun. This practice has been successfully applied, for example, in *Tour Bois Le Prêtre*, a tower in the suburbs of Paris. *Source* The author

and new balconies—from a gloomy accommodation, limited by rigid panels with minimal openings, to the overall view of the city (Fig. 6.6).

6.3.2 Strategies for Sustainable Urban District Redevelopment

The problems in post-war neighbourhoods are complex and are composed of multiple factors at multiple levels. The main categories of problems relate to (Van Beckhoven, Bolt & Van Kempen, 2005):

- housing stock,
- management,
- residents.

The aim of regeneration is to transform deprived neighbourhoods into attractive places that are physically, economically, socially and environmentally sustainable. The Ecopolis model introduced by Duijvestein in 2004 (Gruis et al., 2006) identifies four main perspectives for a sustainable integrated approach (the four Ps):

- ***Planet***, that is the environmental quality,
- ***Project***, which means spatial quality,
- ***Profit/Prosperity***, the economic sustainability,
- ***People***, the social quality and welfare.

The complexity of neighbourhood redevelopment requires a deep knowledge of the dynamics that led to degradation and social crises. The conditions that characterise working-class neighbourhoods differ in their history, location, and emergencies that were not addressed in a timely manner. Social change, immigration and an increasingly complex society are just some of the common aspects that determine degradation. Often some deficiencies were present at the beginning; now it is necessary to identify the most important actions that can trigger a regeneration process. It is important to take responsibility, not only for the administrators, but also for the residents themselves, who must take a dominant and proactive role, as they are responsible for many of the problems that affect the community.

The following are some general aspects that should be considered.

The role of community spaces—Community spaces, which should be available as places for activities, can facilitate spontaneous or guided gatherings of residents. Typical examples include neighbourhood workshops introduced in some redevelopment projects, places dedicated to listening and activating the recreation process, and then neighbourhood management. The size of these spaces can be extremely variable and distributed: they can occupy empty and unused spaces or require integration and new construction. A basis for this is certainly the need for a flexible and variable response over time, able to transform for immediate needs and events, rather than new initiatives that cannot be prevent at an early stage.

Relationships and open spaces—Green spaces are all too often an uncultivated and deserted land. In the rehabilitation of particularly degraded suburbs with considerable social problems, which was carried out in Germany and the United Kingdom elsewhere, for example, the initial strategy was to redesign green spaces, create equipped gardens and create a living space for the district and its inhabitants that had not existed before. The active participation in the maintenance of these spaces, strictly managed by the residents, has helped to stop the process of vandalism and degradation that used to characterise them.

The quality of the green spaces is proportionate to a series of factors such as size, boundaries and functions, which are provided by play and relaxation equipment, and, of course, care and maintenance. The examples designed by Stefan Forster for the redevelopment of buildings in a district of the former GDR are based on the assumption that it is not possible to set a redevelopment process in motion without substantially rethinking open spaces. Forster redesigned all the green spaces according to a hierarchy ranging from private gardens to building-related collective zones and public zones. Paradoxically, after fractioning the space that previously belonged entirely to the district, he has increased the environmental quality, showing that it is not the surface extension that makes an open space perfect.

Mix of housing types and users—It is well known that the population living in the suburbs is extremely diverse, so the first objective is to redefine housing to meet new needs. However, this action alone is unlikely to lead to a process of social regeneration. New forms of housing need to introduce a diversified mix of users, often with short-term contracts, in order to trigger a process of continuous renewal.

New functions—New functions can be introduced that are associated with non-residential areas and can activate a process of renewal of a periphery. To this end, experimental interventions have been carried out that, starting from a conscious analysis of the deficiencies and opportunities in the specific territory, seek to identify functions that represent an opportunity for the inhabitants of the neighbourhood and are also capable of reversing the flows and bringing the "city" to the periphery through work spaces, art, culture and events.

Add or reduce to change the image—The uniformity and repetitiveness of buildings, regardless of their social purpose, create a state of indifference to collective spaces. Perceiving a sense of belonging and recognising the uniqueness of one's space as a whole are important values for residents. Typical suburban social housing neighbourhoods, especially those built from the 1950s onwards, are characterised by a lack of references, the constant repetition of building modules, façades, colours and materials, leading to rejection, degradation and an increase in the number of existing distresses.

The project mentioned above, designed by Stefan Forster for the Leinefelde district of the former GDR, synthesises approaches at different scales (Fig. 6.7). Forster not only analyses new building needs and performance deficits, but also examines the mistakes made in large urban districts that led to rejection and social decline. The theme of urban regeneration is approached starting from the public space, with the reordering of buildings, the hierarchy of outdoor spaces and the recognisability and differentiation of architecture seen as essential. A mixture of low-rise, high-rise, terraced and tower houses, as well as different forms of housing, allows not only for the satisfaction of needs but also for the creation of a liveable urban environment.

Leinefelde presents a veritable catalogue of possible strategies for rehabilitating degraded social quarters. Forster has used various measures to transform monotonous blocks into architecturally high-quality solutions. The most important principles are:

- architectural differentiation of buildings,
- open space hierarchy, from public to semi-public to private,

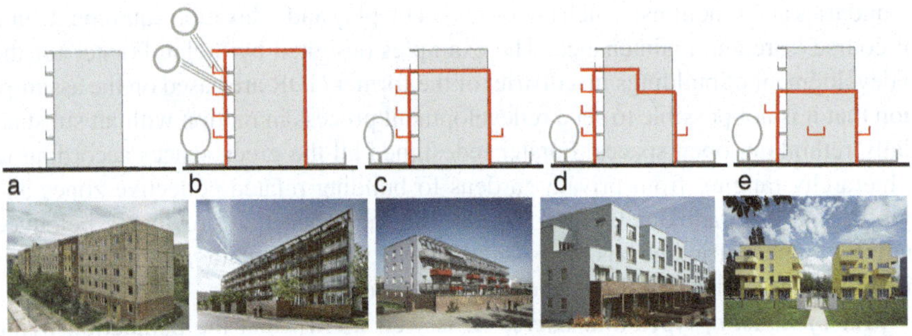

Fig. 6.7 The schematics and photographs show various solutions for blocks developed over a period of about ten years, from the first solution (**b**), where the main objective was energy renovation, new living spaces and private gardens for the ground floor apartments, to the extreme idea for the 180-m block with the creation of a series of separate buildings connected by a continuous wall, reminiscent of the old volume (**e**). *Source* Schemes by the author. *Photo Credits* S. Forster Architect, Photographer Jean-Luc Valentin

- block size reduction through volume deduction,
- volume articulation through ground floor and roof solutions,
- typological diversity to appeal to different population groups.

The projects demonstrate a range of solutions that address the inadequacies of the buildings and improve the context and living conditions in the neighbourhood—solutions that make it possible, if necessary, to guarantee that residents remain in their homes during construction by introducing innovations in the building envelope; or the re-proportioning of the block size when it is possible to temporarily move residents into empty houses.

6.4 Conclusion

The apparent obsolescence of many social housing neighbourhoods that characterise Europe's industrial cities shows that the necessary energy efficiency improvements can be a real opportunity to rethink housing models and adapt housing to new needs.

It is well known that most building envelope interventions and three-dimensional solutions can be an example of how improving spatial quality and living conditions go hand in hand with energy efficiency and resource savings.

Stefan Forster has shown that, in addition to energy efficiency and building adaptation, a rethinking of existing residential neighbourhoods on the periphery of major European cities is also necessary. The quality of the neighbourhood is defined by the balance between the quality of the buildings and the collective spaces.

Acknowledgements This chapter summarises some of the results of various research projects conducted by the author: • "Strategie progettuali e strumenti di valutazione per interventi di edilizia residenziale a basso consumo energetico mediante tecniche costruttive ed impianti ad elevato contenuto innovativo" [Design strategies and assessment tools for low-energy residential buildings using innovative construction techniques and systems with high innovative content], funded by MIUR (Italian Ministry of University and Research), 2004 • "Propensione dell'edilizia residenziale pubblica a Milano ad essere recuperate secondo i nuovi modelli dell'abitare" [Propensity to renovate social housing in Milan according to new housing models], funded by ALER (Lombard Residential Building Company), 2008. • "EASEE, Envelope Approach to improve Sustainability and Energy efficiency in Existing multi-storey multi owner residential buildings", FP7 EU program, 2012—2016. • "Edifici a Zero Consumo Energetico in Distretti Urbani Intelligenti" Zero Energy Consumption Buildings in Smart Urban Districts], financed by MIUR (Italian Ministry of University and Research), 2016.

References

CECODHAS Housing Europe Report. (2011). The place for inhabitants, Residents' participation in creating sustainable housing & neighbourhoods.

Costarelli, I., Kleinhans, R., & Mugnano, S. (2019). Reframing social mix in affordable housing initiatives in Italy and in the Netherlands. Closing the gap between discourses and practices. *Cities, 90.*

Eurostat. (2017a). People in the EU—statistics on housing conditions. https://ec.europa.eu/eurostat/statistics-explained/pdfscache/41898.pdf.

Eurostat. (2017b). People in the EU—statistics on household and family structures. https://ec.europa.eu/eurostat/statistics-explained/pdfscache/41897.pdf.

Eurostat. (2019a). Energy consumption in households. https://ec.europa.eu/eurostat/statistics-explained/pdfscache/58200.pdf.

Eurostat. (2019b). Population and Population change statistics. https://ec.europa.eu/eurostat/statistics-explained/pdfscache/1787.pdf.

Eurostat. (2019c). Young people—housing conditions. https://ec.europa.eu/eurostat/statistics-explained/pdfscache/46039.pdf).

Eurostat. (2019d). Europe 2020 indicators—Poverty and social exclusion. (https://ec.europa.eu/eurostat/statistics-explained/pdfscache/29306.pdf.

Eurostat. (2020). Living condition in Europe—housing quality. https://ec.europa.eu/eurostat/statistics-explained/pdfscache/67027.pdf.

Grecchi, M. (2008). *Il recupero delle periferie urbane. Da emergenza a risorsa strategica per la rivitalizzazione delle metropoli* [*Renovation of urban suburbs. From emergency to strategic resource for the regeneration of metropolis*]. Maggioli Publisher.

Grecchi, M., Malighetti, L.E., & Iannaccone, G. (2012). Integrated design strategies for refurbishment of large social housing complexes in European suburbs. In *Proceedings of the XXXVIII IAHS World Congress April 16–19, 2012*, Istanbul, Turkey.

Gruis, V., Visscher, H., & Kleinhans, R. (2006). *Sustainable neighbourhood transformation, sustainable urban areas* (Vol. 11). IOS Press.

Konstantinou, T. (2014). Façade refurbishment toolbox. Supporting the design of residential Energy upgrades. *Architecture and the Built Environment, 9.* (TU-Delft).

Scuderi, G. (2019). Retrofit of residential buildings in Europe. *Designs, 3, 8.* www.mdpi.com/journal/designs.

Van Beckhoven, E., Bolt, G., & Van Kempen, R. (2005). *Theories of neighbourhood change and neighbourhood decline: Their significance for post-WWII large housing estates.* Paper for the ENHR-conference "Housing in Europe: New Challenges and Innovations in Tomorrow's Cities", Reykjavik, 29 June–2 July 2005.

Van Kempen, R., & Bolt, G. (2009). Social cohesion, social mix, and urban policies in The Netherlands. *Journal of Housing and the Built Environment, 24*(4).

The manufacturer's authorised representative in the EU is Springer Nature Customer Service Centre GmbH, Europaplatz 3, 69115 Heidelberg, Germany. If you have any concerns regarding our products, please contact ProductSafety@springernature.com

Printed and bound by CPI Group (UK) Ltd, Croydon, CR0 4YY
30/09/2026
02225261-0008